THE CREATIVE EDGE

Fostering Innovation Where You Work

WILLIAM C. MILLER

Addison-Wesley Publishing Company, Inc.

Reading, Massachusetts • Menlo Park, California
Don Mills, Ontario • Wokingham, England • Amsterdam • Bonn
Sydney • Singapore • Tokyo • Madrid • Bogotá
Santiago • San Juan

Grateful acknowledgment is made to the following for permission to use copyrighted material:

Addison-Wesley Publishing Company, Inc.: For an excerpt from James L. Adams. *Conceptual Blockbusting.* Copyright © 1986 by Addison-Wesley, Reading, Massachusetts. Pg. 4 (quote). Reprinted with permission.

American Management Association: For an excerpt from *Creative Planning Throughout the Organization.* by Jim Bandrowski. Copyright © 1986 by American Management Association. Reprinted with permission.

continued on facing page

Library of Congress Cataloging-in-Publication Data

Miller, William C. (William Cox), 1948–
  The creative edge.

  Includes index.
  1. Creative ability in business. I. Title.
HD53.M55 1986   650.1   86-14097
ISBN 0-201-15045-X

Cover Design by Steve Snider
Text design by Laura Fredericks
Set in 10 point Palatino by Compset, Inc. Beverly, MA

ABCDEFGHIJ-DO-8987
First printing, January 1987

Dedication

This book is dedicated to the divine spirit we all share; to my wife Sue, with whom I have learned and shared the deepest love, peace, wisdom, and integrity I have known; and to a society beyond scarcity and separation, a world that profits each person materially and spiritually.

Contents

Acknowledgments

My special thanks to:

Joe McPherson, who has contributed so much to my appreciation of the field of creativity and innovation; Darryl Conner, for influencing my thinking in many ways;

Betty Smith and Terry Pearce for helping to shape the book; Ron Skellenger and Scott Shershow, my editor at Addison-Wesley, for their valuable and detailed editing comments;

Deborah Warren and Si Goodwin, of the production staff at Addison-Wesley; Sharon Sharp, copyeditor; Laura Fredericks, designer; and Steve Snider, cover designer, for their creative contributions;

Dave Keaton and Bill Dowdy for their support at SRI; Juanita Brown, Sharon Jeffrey-Lehrer, Cathy DeForest, Brooke Warrick, Roger Harrison, Kare Anderson, and Ron Richards for special contributions and thoughts;

Dennis Jaffe, Newt Gingrich, Rosabeth Moss Kanter, Bob Rosenfeld, Ed Glassman, Crister Ahlstrom, Ryan Petty, Bill Galt, Michael Ray, Mat Juechter, Bill Millette, Bob Larsen, Sushil Bahla and others for their feedback on my manuscript; the many corporate managers who helped formulate examples for the text;

Lee Nelson of Illusion Industries in Oakland, California for his wonderful images for CREATIVE, SPIRITED and APPEARE;

Sue Miller, for her invaluable and tireless support, and in particular for sponsoring creativity gatherings around the country.

Teddy Olwyler, Rich Laine, Jay Howell, Carol Shaw, Dennis Jaffe, Linda Plott, Ron Richards, Ron Howard, Tony Adkins, Hulki Aldikacti, Shirley Adkins, Phyllis Lawson, Barbara Pelletier, Trena Parsons, Bob Glasar, John Gooden, Bruce Beron, John Maher, Diane Wexler, Stan Furrow, Pat Weedeman, Claire Cohn, Cynthia Scott, Terry Pearce, Tom Ucko, Allison Stevens, Tom Green, and all the others who shared with me their hearts and their stories about their creativity at work.

Foreword

Dr. Dennis Jaffe, President
Association of Humanistic Psychology (1985–1986)

From often painful experience, we have an image of organizations as straight-jackets, muzzling the human spirit and dampening the fires of excitement, creativity, improvisation, and spontaneous community. In the past, organization was seen as inseparable from bureaucracy, which emphasized the rational over the exceptional, the predictable over the wondrous, and the routine over the new possibility.

But the Freudian notion that we need organizations and cultures to limit our instincts and channel us by force into useful endeavors is not working so well for us. Organizations today cannot continue to do things as they have always done them. The premium is now on discovery, initiative, and continual re-creation as an everyday facet of organizational life.

The new realities demand new styles and forms of organization and demand new skills from the people who live within them. As we move into this new territory, we need some maps, and we need to learn to use new types of tools.

Bill Miller has created a Baedecker of this new world, a guide book full of incredible richness—stories, fables, and practical tips for liberating the human spirit to re-create the organization. What is most special about Bill's work is that he has successfully created a model of creativity and innovation that spans the gaps between *the individual* person expanding his or her potential in their work, and *the team, work group,* and *organization* that wants to liberate the creativity of many people together without damaging the "higher" systems of production, productivity, and control.

The book is full of stories about individuals in organizations—not lone wolves, and not abstract systems without real people. The stories do not give cookbook-type solutions or impossible-to-emulate heroes but have the effect of leading the reader to think creatively about what is possible in his or her work situation.

The Creative Edge makes the very special, sensitive, and personal work of Bill Miller available to a broader audience. Bill is a consultant of deep experience and rich creativity. He is able to span many disciplines and to integrate frameworks ranging from humanistic psychology to large system dynamics into his presentation. But his perspective is always clear—his aim is to deal with the real problems and crisis areas facing organizations today with the kind of concrete, specific, practical ideas and possibilities that will help harness the vast inner resources of the people who make up an organization to meet the challenges of innovation, increased productivity, and concern for human well-being.

Bill Miller's book represents a kind of corporate new wave. He takes up some of the principles and themes of the new "excellence" literature, which so far have been either too abstract or too purely inspirational, and makes them useful to managers and whole companies. He talks about real human-scale achievements, but he does so with a deep awareness that large, complex organizations are difficult systems, with their own sets of rules and constraints.

xiii

Foreword

Newt Gingrich, Congressman
U.S. House of Representatives (R–GA)

I first met Bill Miller in 1973 when he was a graduate student at West Georgia College studying Humanistic Psychology. At the time I was a History and Environmental Studies professor, trying to blend the lessons of the past with the opportunities of the future on a remarkably diverse college campus. Bill, as a Californian imbued with modernity, was living at the edge of the future. He was interested in better understanding a small town in rural Georgia. The differences in rhythms and life-styles of the large cities of the West Coast and the smaller, more traditional town of the rural Southeast were compelling.

In the beginning I was fascinated with Bill's ability to take ideas from a variety of sources and teach new insights, new approaches, and new lessons to people who couldn't quite envision the world beyond their own psychological and historical perspectives. Over the last thirteen years I've watched him grow in sophistication and understanding, and importantly, translate that growth into expressions of practical policy.

Bill has developed a vision of creativity and the process of being creative. That process is useful to anyone trying to invent a more human, progressive, and exciting twenty-first century. In *The Creative Edge*, Bill Miller synthesizes his experiences, his wide range of study, and a set of practical guidelines that prove remarkably stimulating and inspiring.

My seven years as a U.S. Congressman have convinced me that the key to America's future is reacquiring the creative edge. Government, business, volunteer associations, and local communities must reestablish an optimistic creativity that helps us solve the challenges of creating a better future. In such an atmosphere, I have no doubt the twenty-first century will be one of freedom and prosperity for everyone. It is in this spirit I urge you to turn the page and begin rebuilding the creative edge.

Introduction

Victor Equipment Company, like many organizations across America, has the admirable habit of regularly conducting employee attitude surveys. As their Corporate Manager of Training and Development in the late 1970s, part of my job was to work with plant managers around the country to analyze their surveys, discover training and development needs, and help develop appropriate programs and budgets.

While the surveys generally showed positive responses to management practices, we needed to explore comments such as, "We don't get paid enough around here," especially when employees might add, "That is, we don't get paid enough to put up with some of the things management does." At the base of many of these responses was the sentiment, "I want to do a good job, if they would just let me do it."

When we pursued the meaning of employee complaints, we consistently discovered that the core desire of employees was to feel that they were making *as important a contribution as they were personally capable of* to a productive, profitable business, *and to be recognized for that contribution.* Bill Hewlett, co-founder of Hewlett-Packard, expresses it this way in "The HP Way"[1]:

> *I feel that in general terms it is the policies and actions that flow from the belief that men and women want to do a good, a creative job, and that if they are provided with the proper environment they will do so. It is the tradition of treating every individual with consideration and respect and recognizing personal achievements.*

A good job . . . a creative job. It is from the creative contributions of people like yourself that organizations grow and flourish. The empowerment of each of us to be all we can be is the foundation on which profitable, successful organizations are built.

Peak-performing organizations are in the business of creativity. They innovate; they take their creative ideas and produce something with them. That's how they renew themselves and how they stay healthy and profitable to serve their customers, clients, stakeholders, and themselves. This is true for major corporations, grass-roots nonprofit services, government agencies, and small businesses. Whether the creativity is expressed in new products and services, new public relations approaches, new programs, or new marketing or merchandising methods, success in the long run depends on creativity and innovation.

Yet creativity is feared; it seems to imply being unrealistic and "off the wall." Innovation is avoided; it requires change, change that must be managed effectively. Both creativity and innovation can be focused to be strategically appropriate and personally fulfilling. Change can be managed to meet the diverse needs of the people who implement it and to work constructively for the organization as a whole.

Innovation and creativity have become, in some cases, complex parts of today's business world that cannot be dismissed, and in others are considered only when short-term, fix-it solutions are required. They are strategic issues affecting every facet of product development, manufacturing, marketing, and sales. New products—or new marketing or sales strategies for that matter—require the collaboration of many disciplines within an organization. They require creativity and innovation every step of the way.

The evidence is mounting that as the nature of the business environment is changing, so also is the nature of management. It now requires the inclusion of formerly "irrelevant" and "impractical" notions. For example:

1. North American Tool and Die (NATD), led by Tom Melon and Garner Beckett, has an amazing track record. Since 1978 its sales have increased fourfold, its profits seven-fold. Its stock appreciated 36 percent in 1980, 1981, and 1982. Its product quality has seen the reject rate drop from 5 percent to 0.1 percent. Its productivity has doubled. Turnover dropped from 27 percent to under 6 percent. Melon and Beckett credit a strategy deliberately designed to build *trust* between employees and owners.

   In the beginning the owners set four objectives: steady the growth of the company, increase profits, share the wealth, and help everyone find *satisfaction and fun in their jobs*. NATD's corporate culture reflects these goals and is a key factor—perhaps THE key—to its overall success.

2. The Olga Company is well known and very successful in the lingerie industry. In 1981, for example, its sales increased about two and one-half times, and its return on invested capital (before taxes) was four times that of other foundation/lingerie firms (and 160 percent of the norm for all apparel firms). From its inception in 1941 its founders committed themselves to recognizing and implementing Judeo-Christian values in their workplace and business activities. One of these values even today is *leadership through service*. Jan Erteszek, Chairman of the Olga Company, says, "It is the manager who, in effect, works for the members of his or her department." In an assembly operation, for example, the manager must see that a climate exists in which employees can grow personally and professionally while maximizing profits.

3. When Jan Carlson became President and CEO of SAS airlines, it was in the red on domestic service and ranked well down the list among European competitors. What did he do? He first turned the organization chart upside down and instituted policies based on management serving the employees who interfaced with customers (more "leadership through service"). "How can we support those employees better?" he asked. He wanted an organization that could do 1,000 things 1 percent better rather than trying to do 1 thing 1,000 percent better. This, and implementing a firm realization of the need to be customer-driven, meant that within two years, SAS was well into the black domestically and became a top European airline.

4. W. L. Gore and Associates, makers of Gore-Tex, has grown very rapidly into a $1 billion company with 5,000 employees and 40 plants and *no* system of job titles, management maps, or formal bosses. A philosophy of "management by volunteer commitment" results in new hires volunteering to work on projects that strike their interest. Their success is based on an understanding that commitment, not authority, is what produces outstanding results in the workplace.

These stories are not just flukes. They are underscored by the extensive research done by writers such as Tom Peters and Bob Waterman, Rosabeth Moss Kanter, Gifford Pinchot, John Naisbitt, Warren Bennis, Peter Drucker, and Charles Garfield. There are principles to follow but no prescriptions to guarantee results.

The greatest challenge to our creativity is to constantly evaluate and adjust how we manage ourselves, our work, our relationships, and our companies to maximize their potential. Each person, each company, and each situation is different. We know comparatively little about these creative adventures we could undertake daily. But I believe that rather than relying on outside experts, we need to look deep within ourselves for solutions. One thing, however, is certain: creativity is the prime source, the taproot, from which many solutions will spring. But what is its nature?

Mike Arons, chairman of the Humanistic Psychology Department at West Georgia College, once began a class on creativity with the question, "What is 'creativity' . . . such that other cultures, even with Latin-based languages, may have no word for it?" He explained that when he received his Ph.D. in creativity from the Sorbonne in Paris, the French language had no such word as *creativité.* There was *ingenuité, creation,* and *originalité,* but no *creativité.* (The word was subsequently recognized by the French government committee that annually updates the French language.)

Mike's question is still a very stimulating one. What is creativity? What do we (or more importantly, you) refer to when using the word *creative*?

This is a key question because whether you see yourself as creative or not depends on your personal definition of "creativity." And seeing yourself as "creative" has a large impact on how you express your inherent potential while living and working creatively.

Charles Minghus, the jazz trumpeter, says that creativity is making the complex simple, as Bach did in his music. There are many semantic debates over whether something is "truly creative." Is it inventive or merely innovative? Evolutionary or revolutionary? Merely novel or a real breakthrough?

The *American Heritage Dictionary* defines creativity as the "ability or power to create things; creating, productive; characterized by originality and expressiveness; imaginative."[2] These four dimensions—creation, productivity, originality, and expressiveness—all suggest that creativity is the full actualization of our human potential. Inspiring and coordinating people as a manager, constructing or assembling a product, satisfying a customer's service needs, writing letters or memos, counseling a student, or determining business strategy can all be either mechanical, lifeless routines or imaginative, productive, expressive, creative activities.

Throughout this book, when *creativity* and *innovation* are both used, the former refers to conceiving of and developing new ideas, and the latter to implementing them. When *creativity* is referred to by itself, it implies both originality (idea generation) and productivity (implementation).

These may not be your definitions. Perhaps you would agree with writer Dorrine Turecamo that "creativity, after all, is imagination in pursuit of excellence,"[3] or perhaps you would say as Justice Stewart did in defining pornography: "I don't know how to define it, but I know it when I see it." But then, creativity and innovation are words that deserve multiple definitions. What *you* mean is what's important. How *you* define it determines whether you acknowledge and express your own inherent creativity.

The Creative Edge aims to help you develop your own productivity, originality, and expressiveness, and that of the other people in your organization. The challenge is the same for all of us: To be all we can be, individually and collectively. The goal is more productive response-able organizations where people make as important a contribution as they are capable of and are recognized for that contribution.

Whatever your experience of creativity in your work, this book is intended for those of you who want a new vision of yourself and your workplace—for those of you who want yourself and your organizations to be more creative, productive, resourceful, motivated, responsible, profitable, and alive and growing. It is for those of you who want to experience the other side of stress and burnout: the creative side that profits everyone.

The Creative Edge is not only addressed to those of you who are managers and R & D technologists, who often have official responsibility for being creative or managing creative and innovative people. This book is intended for any of you who wish to enhance yourself, your organization, and your society. All of us are managers of our own work, and all of us can be involved in the "R & D" of making our organizations more effective, accomplishing goals, and contributing to society.

The Creative Edge offers you an introduction to the knowledge, skills, and attitudes needed by people throughout an organization—at all levels of the hierarchy—for fostering strategically appropriate innovation and creativity. It is a practical guide to reframing your view of creativity, developing your inherent creative skills, and promoting creativity in others—no matter what your job is.

This book is not an academic research treatise on creativity, nor is it a psychological profile of the creative person. It definitely is not the end-all and be-all on the subject of fostering innovation. It doesn't discuss product and innovation life cycles, R & D funding, and a host of important topics for a wide variety of ever-changing business decisions.

Instead, *The Creative Edge* focuses on the environment, the context, in which those decisions must be made.

To accomplish this goal of fostering creativity and innovation, the book is organized into four sections:

1. *Fostering Creativity and Innovation from Anywhere in the Hierarchy* invites you to expand your notions of what it means to foster creativity and innovation where you work.

2. *Individuals Working Creatively* gives you specific models and methods for expressing more of your individual creativity.

3. *Groups Working Creatively* gives you specific models and methods for promoting creativity and managing change in groups where you work.

4. *Of Profits and Prophets* develops ideas for living and working in harmony with the goals of both the achievers and the societally conscious in this country.

Examples and stories used in this book are not all from the "excellent" companies or the "best companies to work for." As we shall explore, you do not need the optimum work environment to be able to assert your creativity.

Also, many key topics covered in this book could have whole books written on them (and perhaps already have). One intent of this book, however, is to integrate such topics into a coherent whole for fostering creativity and innovation.

In sum, this book will help you view your creative abilities in a new way and to explore specific ways in which you can exercise your creative potential while promoting creativity in others. As a result, you will develop a more powerful expression of your talents and capabilities, your love, and your wisdom as they apply to your work life.

The Creative Edge

PART I

Fostering Creativity and Innovation from Anywhere in the Hierarchy

"You know what I'd like to do, Caslow? I'd like to create a far-reaching, innovative program that will open a lot of channels, offer great opportunities, link up with all kinds of things, and enable something or other to happen. Any ideas?"

The best way to predict the future is to invent it.

—*Alan Kay,*
A Fellow of Apple Computer

CHAPTER 1

You and Your Creativity

Your Creative Edge

Have you ever had an exciting project where you were taking a novel idea and really putting it to work only to run into one of the Charles Duells of the world (the director of the U.S. Patent Office in 1899, who said, "Everything that can be invented has been invented")? Do you work with a Harry Warner (the president of Warner Brothers who said in 1927, "Who the hell wants to hear actors talk")?

Have *you* ever been an Erasmus Wilson, the Oxford University professor who said in 1878, "With regard to the electric light, much has been said for and against it, but I think I may say without contradiction that when the Paris Exhibition closes, electric light will close with it, and no more will be heard of it"?

There is perhaps no greater celebration of life than our creativity, *especially* at work. To profit and prosper, our organizations face no greater challenge than to promote and guide that creativity.

Consider, for example, the case of a large manufacturing company whose V.P. of marketing approached its president and said that he was worried about having competitive products to market in the coming five to ten years. Based on some research, the V.P. believed that their competitors were gaining a creative edge with more innovative products, spelling trouble for future profits and market share.

As a result, the company called a conference of its top executives—marketing, planning, finance, and operations—to explore what was and what wasn't working in their climate for creativity and innovation. They asked themselves:

- How do we feel about our current quantity and quality of activity for developing and marketing new products?

- How can we build on and reinforce the factors that have contributed to our past success?

4

- How can we remove or bypass barriers?

- How can we set goals for and measure appropriate new product development activity?

- For what markets do we most need to improve our new product development process?

They gained many insights, not the least of which was that their culture did not reinforce the need for people to be creative and innovative. Short-term financial performance was more important than long-term innovation. Creative new products were something to be bought up through corporate acquisition rather than developed through human and technological resources. Creative people were not recognized or rewarded and in some cases were actively discouraged for "rocking the boat."

This story illustrates our critical *business* needs for fostering both creativity (the birth of imaginative new ideas) and innovation (the transformation of those new ideas into tangible, practical products, services, or business practices). Our organizations, in both the public and private sectors, face a common need to innovate, whether to replace maturing revenue streams, to stimulate a market/client-driven (rather than technology/product-driven) culture, to keep pace with rapidly changing markets, or to improve morale within existing mature businesses.

Our *human* needs for creativity and innovation are equally compelling. Some of us have responded to a lack of creativity in our work with strong feelings of frustration and anger or with "burnout." Others have known the vibrancy of innovation through the satisfaction of being productive and meeting organizational and personal goals.

As symbolized by the edges of this pyramid, there are four ways in which you can experience a creative edge where you work:

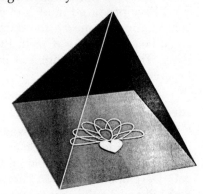

1. Promoting your organization's leadership in its chosen products or services

2. Challenging the frontiers of marketing, sales, or technology

3. Taking an active part in change

4. Exploring and developing your own inner inspiration and creativity

The design on the base of the pyramid represents the source and human values from which full creativity emerges.

A creative edge does not imply a win-lose relationship. It does, however, require developing yourself, your co-workers, the climate for innovation, and the creative process to their fullest. You can make a difference where you work both to business and human needs. You can experience a creative edge in your work life. First, you must recognize your potential to act creatively.

In The Creative Spotlight: You

> ◖ *It seems to me that where I'm creative in my work right now is really insignificant—things like the way I dress. I simply don't want to slide into being a regular librarian. I find we get very bogged down—we're understaffed and underbudgeted—and nothing gets done. Sometimes I feel a little outrageous and do very small things a little differently. I had an outrageous little party to celebrate the finishing of a project. I think in this particular setting, such things aren't usually done.*
>
> *Teddy, librarian* ◗

When asked, "Are you creative?" many of us might answer, "Sometimes . . . It depends on what I'm doing." If asked, "Are you creative at work?" our answers may sound more like, "Here and there," or "Rarely, but sometimes I have my moments." Sadly, all too many might say, "In my job, I don't get the chance to be creative." Some of us, fortunately, answer unequivocally, "Yes."

I know that some of the time I do not recognize when I'm really being creative, and I overlook opportunities to be creative. As a result, I experience a whole range of feelings—excitement, frustration, joy, disappointment, enthusiasm, anger—and my productivity at work is often mirrored by these feelings.

I have produced more and enjoyed myself more at work as I have come to understand all the different times when I've been creative or could be creative. For myself, I've identified seven arenas for my creativity. Think back to different moments in *your* life and see if you've experienced them as well.

Idea Creativity

Can you recall a time when you had an insight into or a solution for a problem? Understood a problem in a new way? Developed the concept for a public relations campaign? Came up with the idea for a new product? Developed a recipe? Put various thoughts together for the first time? Phrased a thought in your own terms?

This realm of new *ideas and concepts,* all involving intuitive or logical thinking, is an important arena for creativity. Creativity has often been applied to thinkers, especially in the disciplines of science, technology, advertising, and philosophy. Even a new understanding of the source of a disagreement or a child's discovery of how to stack toy rings belong in this category of creativity.

> ◖ *Somebody came to me recently with the problem of making silicon nitride powders. We can make silicon nitride fibers, but could we make very fine powders? We came up with some solutions that we thought could do it, but neither of us were satisfied with them.*
>
> *About a month later an idea popped into my head which I think solves the entire thing, but I don't know how I reached that conclusion . . . Instead of making polymers beforehand and spraying them and heating them, (the idea is) to create the polymers as you spray them.*
>
> *Rick, lab director in organo-metallic chemistry* ◗

Q In what ways have you been creative through new thoughts or ideas?

Material Creativity

Can you recall a time when you wrote a report or book or paper that you especially liked? Designed or produced an advertisement? Engineered a piece of machinery or electronics? Took a photograph? Composed a piece of music? Drew or painted a picture? Designed a costume or other piece of clothing?

All of these involve creating and producing something in the *material, sensory world*. Creativity has often been applied to material works of art and various new consumer products, which fall in this category. Whether engineering a product or a report at work, these involve expressing yourself by making something concrete and real for other people to share *repeatedly* and appreciate by seeing, hearing, touching, smelling, or tasting.

> ◖ *When I'm being creative, I feel like I'm in a flow, and the results I'm producing are superior—for instance, in writing a business report. Sometimes I sit down and think, "Oh, this is not only a logical way of looking at this, but it's also a way I haven't thought of before—a new perspective" or, "This is a way to present this in the best possible way." And then when I get finished it doesn't look like just another business report.*
>
> *The report comes more from my heart than my head, but it's full of facts: very cost-conscious and bottom line, with benefits and costs involved. Otherwise, even though it could be*

*technically well put together, with excellent grammar, it still
wouldn't have that spark to it.*

Sue, project manager of computer systems in a bank ◖

Q In what other ways have you been creative by producing
something material or sensory for others to experience?

Spontaneous Creativity

Can you recall a time when you gave a speech "off the cuff"? Sang a
song in a new way or improvised music "ad lib"? Made an inspiring
play in sports (a special move while running with a football or adjusting
to the bounce of a tennis ball, for example)? Responded spontaneously
in a meeting or a work conversation with a sincere plea or justification
for an idea?

All of these instances involve a sense of spontaneity, a moment not
to be repeated easily. *Spontaneous* creativity is often experienced as being
at our best, purposefully expressing our competence and confidence in
the moment.

◖ *I really trust spontaneity to carry some creativity. I'll do things
on the spur of the moment knowing that the thing that makes it
is the creativity. At first, that's pretty terrifying. At the same
time, once you begin to love that process, you become committed
to it.*

*For example, I did a lot of sales work, and you certainly
can play a sales presentation very safely. I've had times when
I've said to a prospect, "This isn't touching you, is it?
Something, some way that I'm presenting this isn't speaking to
you. So let's talk about that, instead of what I'm selling." That
is an example of being willing to risk and see what happens.
More times than not, it works because people are dying for that
real stuff. In risk is creativity.*

Jay, community relations director for a college ◖

Q In what other ways have you been creative through new,
spontaneous expressions of yourself?

Event Creativity

Can you recall a time when you directed a meeting at work? Developed
new procedures for getting work done? Planned a wedding or holiday

dinner? Devised a way to gain consensus in a strategy discussion? Organized a picnic?

Sometimes creativity has been applied to *events* or *processes* that have been uplifting or moving experiences. The fluid mixture and flow of decor, people, sequence of happenings, and background are part of the process of creating events, whether just for yourself or for others, too.

Or perhaps you once said, "I want such and such in my life," and you eventually got it. *Circumstances* that we wish for and find ways to fulfill make up another aspect of this type of creativity.

> ▰ *I had the idea for the conference and let it sit awhile. The process took two months from the time we said it was a "go." Picking the speakers was more intuitive. We worked the two months on the specialty sessions, calling people, getting leads—kind of like a detective story, running into people and asking, 'Who would be good for that kind of session?' And I visited or called or wrote almost every chief executive officer, and also sold the blocks of tickets. I would say half the conference registration came from that. It's one of those things where you do it once in a certain way, and then if you have to do a different one next year, you do it in a different way.*
>
> *Carol, assistant dean of engineering at a university* ▰

Q In what other ways have you been creative with special events or circumstances in your life?

Organization Creativity

Can you recall a time when you put together a new project/work team? Started your own business? Joined a political party or a church? Changed the policies and rules of a work group? Developed a new system for getting things done?

As people, we create organizations and social movements—purposeful "families" and groups. This arena also involves creating "structures" and rules to guide social interactions and tasks over a sustained period of time.

> ▰ *One of the biggest challenges and creative moments I've had was in the sixties. A bunch of us were working in a mental health center and just decided we were going to start our own facility. My whole life was marked by the fact that these challenges were so immense. We realized, "Gee, we don't know anything about managing and how to organize. How do we make people do*

things? How do we get results? How do we set things up? What do we do with the money? How do we do books?"

The way in which we learned showed how quickly twenty-two-year-olds can learn. We were all supposed to be just case workers, at the lowest level you could possibly be in the mental health system.

Dennis, clinical psychologist and consultant ◀

Q In what other ways have you been creative in organizing and maintaining ongoing groups in your life?

Relationship Creativity

Can you recall a time when you said to someone (or yourself), "I like the way you handled that situation. You dealt with that person very well"? A time when you resolved a difficult conflict by trying a new approach with the person? A time when you had a romantic evening? A time when you really felt that a co-worker was also your supporter and friend?

Relationship creativity focuses on how you develop collaboration, co-operation, and "win-win" relationships (i.e., those in which both parties feel good about the interactions). Whether in our jobs or in our homes, what we make of our relationships is a true signal of the values that underlie our creativity. We have many degrees of freedom in how we relate to others in any given circumstance. How we finally express ourselves—in words, actions, and even feelings—is a creative act.

▶ *You might not think that a person in real estate management, dealing mostly with tenants and money, is creative. It's a hard business! After thinking about it, my creativity is relational—working out conflicts and problems where no previous pat answers existed. My process is to let conflicts arise, not smother them. They're hard at first to deal with—usually unpleasant—but out of expressions of feelings, a compromise can always be found.*

In our business, and with a small staff, it is not cost-effective to spend a lot of time on a problem and find the perfect solution. You find something that works, do it, and keep moving on. You'll find out soon enough if it doesn't work, and then you try something else.

Linda, real estate manager ◀

Q At what other times have you been creative in how you related to others?

Inner Creativity

How often have you experienced, or witnessed, a "change of heart"—a change of perspective, of feeling, or of attitude where the situation didn't change, but the person's point of view did? Have you ever heard someone say, "Cheer up. Don't be so sad" or "Cool down that hot temper of yours"?

Statements like these point out that we can change our inner emotions even while the outer situation remains the same. We can create different *"inner"* experiences of events. This type of creativity can be nurtured in ways that encourage inner strength and openness rather than a control or suppression of feelings.

The term *creativity* has not often been applied to this arena of inner experience, perhaps because our perceptions so often focus on external objects (for example, the creativity of a sculpture), or because to many the act of creativity implies some sense of will and control, whereas feelings and thoughts often seem to come to us without control ("If I could stop feeling sad, I would!").

If ideas and insights are internal happenings that are considered creative, then other inner experiences can likewise be considered creative. I think of both thoughts and feelings as "content" or "internal products" that we can eventually create, so that a new inner experience spontaneously occurs without a change in situation.

Your inner experience is the most fundamental arena of creativity. Opening yourself to acting creatively through generating new ideas, producing something in the material world, spontaneously expressing yourself, shaping events and circumstances, organizing group life, and developing relationships is enhanced by exercising your inner creativity.

Even managing stress and preventing burnout depend greatly on exercising creativity at this level. Seeing your own creativity in a new light—reframing how you perceive your creativity and expressiveness—is an experience of inner creativity.

> ▶ *I've been a president of a company, and I feel rapport and kinship with the executive world; but I used to have quite a lot of anxiety preparing to speak to groups of executives. I would put in mammoth amounts of preparation, yet my ratings were mediocre.*
>
> *Then I was asked to give an important speech. And I began working on that speech, and I did a lot of things I had never done before. I resolved that I would have absolutely no limits. As I faced the audience of 200 people, I smiled at them, they smiled at me, and I could literally feel the remnants of low-level anxiety draining out through my legs. I felt nothing but excitement, confidence, delight at being there.*
>
> *Rather than being stiff, tense, lacking animation, lacking variety, no drama . . . there I was, audience with me, live*

interaction, a lot of humor, a lot of pauses, a lot of drama. I said to myself, intellectually, "Hey, the body-felt sense of excitement and anxiety are just about indistinguishable. It's your interpretation."

Ron, marketing consultant and professional speaker ◼

Q   In what other ways have you already experienced inner creativity?

Creativity, therefore, has a rich set of expressions. It means more than just being imaginative or productive. However, the ultimate definition of creativity in your life is your personal one. "Working creatively" might just evolve for you over time. Although you may wish to add other arenas where you are creative, these seven—*ideas* and concepts, *material* world, *spontaneous* happenings, *events* and circumstances, *organization* of purposeful groups, *relationships* with warmth, and *inner* experience—are a place to start in further developing your creativity.

Productivity and Creativity

In spring of 1985 ABC News's Peter Jennings and Holocaust scholar and author Elie Wiesel had breakfast in Krakow, Poland, lunch in Auschwitz, tea in London, and dinner in New York. The next day Jennings was back anchoring the newscast. "Tired?" he was asked. "It's the days when you haven't made a really creative contribution when you're tired," he replied.

We are all naturally creative. Without creative expression in my own work life, I personally feel a subtle (or not-so-subtle) flabbiness. My creativity "muscles" want to be expressed, stretched, and exercised; and I believe many of us often feel this way. I have never met an executive, manager, professional, clerical worker, factory worker, or anyone else who did not want his or her job or organization to be filled with trust, collaboration, productivity, and a general aura of success.

This has important implications for our personal and group productivity, which in turn affects how we foster innovation and creativity. The need for day-to-day productivity is often the biggest barrier to encouraging creativity and change.

Productivity is generally defined as a family of ratios of output to input: the relationship between what's produced to what was first brought in. Productivity is rarely an organization's primary objective. It can, however, be a means to achieving many of its goals in a multitude of categories: finance, products and services, employee development, and community service, to name just a few. Productivity planning is best focused on reaching these types of objectives.

When productivity is considered an organization's primary goal, its employees can begin to feel either directionless or myopic. The end result is the same: employees working hard and doing something well but not understanding the meaning behind doing it. Even when creativity is focused on productive ways to meet financial goals, for example, this lack of meaning can persist. This is because finance at its core is simply the exchange of value, and value is based on our human values.* Financial goals have meaning only for what they really symbolize: controlling events, pleasing others (including stockholders) and thus being honored and respected, feeling self-worth, developing an environment for belonging, experiencing culture, and so on.

Productivity and creativity are closely linked, both in definition and in human motivation. When we are our most creative, concentrated and expressive selves, we are also most productive. Creativity is rarely our primary objective, yet it helps us attain our objectives and goals.

To experience contributing to the world, we must produce a product or service. We must have something to exchange. Personal productivity is based on our talents and is directed towards a satisfactory exchange of value (an exchange based on what we each "value"). This exchange first occurs in organizations *not* between the customer and business but between an organization and its employees. It helps dramatically when the organization supplies an environment that supports creative, collaborative effort.

Productivity improvement can be viewed as simply innovation turned "inwards" rather than towards new products or marketing. While intending to build productivity executives conduct employee attitude surveys, implement quality circles, reorganize departments, move staff around, install training and development programs, formulate new budgets, and devise new improvement themes. But like many others, whether top executives or minimum-wage workers, I've become frustrated when I've felt the day-to-day pressures for production overwhelm the intentions for improvement.

When people say, "We need to be more creative around here" or "How can we foster more innovation?" there are two common reactions: "You're right, but what are the odds of changing things here?" or "We're too busy with all the other improvement programs and making this quarter's goals." With a sigh of powerlessness, we all too often quickly sink back into old routines with renewed frustration and "vigorous ap-

---

* Even a self-sufficient farmer on his or her own land has an economy going: if he or she does not exchange the value of renurturing the soil properly, the land will eventually cease exchanging the value of its crops (witness the cotton fields of the 1800s). When Coolidge said, "The business of America is business," he just as well could have said that the business of America is value(s).

athy." And we all find ourselves thinking, "If only *they* would be different."

Many people *want* to make a difference in how their organizations operate but wonder *how* to have an impact. Promoting both creativity and productivity requires more than good intentions. It requires an inner sense of personal power; an understanding of how the organizational climate for innovation is fostered; the cultivation of skills specific to this goal; and patience and persistence in allowing a group of people to do what it has to do to transform itself. It requires management from every level in the hierarchy!

Productivity and the New Work Ethic

"Managing productivity" has become a well-accepted notion, but "managing creativity" appears to some people to be a paradoxical concept. Management implies control, and control seems to be the opposite of creativity. But when "management" becomes "leadership," creativity can be managed without stifling it.

Some managers still practice the philosophy that people can be controlled or manipulated into thinking, feeling, or acting in specific motivated ways; this is mechanistic thinking that is proving less valid in today's work environment. The design of idiot-proof jobs—jobs that do not inspire employees and leave little room for creativity—is another consequence of this "control" mentality. Many people feel alienation, isolation, afflictive stress, and even burnout from working where this mechanistic management is still attempted.

By contrast, leaders *empower* people rather than motivate them. They supply "environmental encouragement": corporate culture within which people can exercise their inherent talents for creative and productive work.

Leaders specify challenging "outcomes" that require creativity to be fulfilled—working in creative new ways to produce creative new outcomes. Their directives emphasize more *what* to produce rather than how to produce it. And even with the question of "what," you are more likely to be consulted in the decision or have a co-responsibility in setting objectives. Akio Morita, president of Sony, says, "If you give people a clear target, you stimulate their creativity . . . You can't make people creative just by telling them to be. You must give them the target." Having to put a new chemical processing plant in operation within a year or having to double revenues in a mature industry within four years can demand creative answers.

I do *not* mean to imply that only senior managers can be leaders in our organizations. Nothing could be farther from the truth. We are all managers of our own lives, and to varying degrees we manage our own work—what we do and how we do it. As consultant Chris Hegarty puts it, we can "manage our boss." He notes that "no matter where you stand on the organizational ladder, you may have the power to transform your relationship with your boss. The responsibility for making it work falls on both parties. To a large extent you have *taught* your boss how to treat you. And, even if you have felt victimized by your boss' personality or methods, you may be able to change the way he or she treats you so that both will benefit."[1]

We all have bosses; even the chief executive officer reports to the board of directors. Leadership works at every level of an organization; it is often the effective smaller work groups that hold an organization together. Understanding how to promote more creativity in our work can assist us in managing our bosses at all levels, for the benefit of all.

> **Q** To what extent do you feel that you spend your time following orders rather than generating your own activity? Are you exercising your creativity as well as you could? What does that say about how you manage your boss?

Leaders spend their time developing vision and strategy, building bridges between organizational units, focusing on leveraging the value of knowledge workers, and building trust.[2] They also invest in and take risks with individuals. Bob Metcalfe, president of 3COM, says, "(Innovation) requires gambling and risk taking. We tell our folks to make at least 10 mistakes a day. If they're not making 10 mistakes a day, they're not trying hard enough."[3] Every false step is an opportunity to learn, and not the end of the world.

Leaders are not just a charismatic few but people who realize that leadership can be learned and practiced. They realize that the old work ethic hasn't died, as some have feared, but is reemerging in a new, more responsible form: the popular movements for participation and involvement reflect a trend of taking *more personal responsibility* for decisions. The newer work ethic strongly values individual responsibility over order following, personal expression (creativity) with our unique talents over fitting in, and creativity over business-as-usual.

The new work ethic comes at a time when flexibility is increasingly required to meet the accelerating pace of change in economic, political, and social realms of our world. As such, it can be seen as a societal survival mechanism that offers us the highest of benefits. It is part-and-parcel for managing in the information age, especially with the growing body of knowledge workers whose brainpower has become the key corporate asset rather than natural resources.

Ironically, this heightened individual responsibility is in tune with America's entrepreneurial traditions but not in tune with management practices in some of our larger business and government institutions. The whole rebirth we are seeing in "intrapreneurship,"[4] business venturing, and creativity as a whole is a sign that the new work ethic is beginning to mature just when we need it.

Organizations of all sizes need to harness and harmonize with this work ethic. By 1990 the majority of our GNP will come from businesses with less than 500 employees. Fortune 1000 corporations are often struggling to match the flexibility and responsiveness that many of these smaller businesses use as their competitive advantage; internal ventures and "intrapreneuring" are several approaches that have been tried—with mixed results. Growing companies are wondering how to sustain their entrepreneurial spirit as they get larger.

Part of the competitive strength of smaller organizations has been explained as "economics of scope," rather than "economics of scale." Yet, I believe, the true advantage of smallness is that individuals can easily feel that what they do makes a real difference! What people in organizations—you and I—must do is foster a climate in which we can experience making that difference and having a creative impact. How to create that climate, in practical terms, is what this book is about.

IN CLOSING . . .

A CEO I know generates such loyalty and teamwork that even though his organization has gone through layoffs and the CEO has personally had to say, "I'm sorry it has to be you," people have said to him, "Is there some way I can come back to work for this company? I like working for you, so if there is any way to do it, at any time, tell me and I'll be here."

This same CEO once addressed a group of his peers—a group of outgoing CEOs that had collectively experienced the wide range of traumatic events and pressures that CEOs deal with—and made this analogy: "We have to keep reaching for truth and beauty. Even if we are in the midst of the muck, we have to think about being like a lotus blossom. Float in the dirty water, but bring beauty out of it." The people he said that to are still repeating that story.

Ron, director of an association of CEOs ◀

I go from being a carpenter to electrician. I see what I need and sit down and do it, and sometimes I just have to throw it away—it doesn't work every time. I try to do everything so it will last. I don't try to do something to get by with it, because you have to come right back and do it again. If it takes four bolts to hold it, I don't put three.

Tony, highway maintenance supervisor ◀

Creativity and innovation require both inner searching and outer action. Within each of us lies the whole world of our potential. If we know how to look and learn, then the key to unlock that potential is in our hands. No one can give us the key to the door except ourselves. Regarding outer action, Peter Drucker states that ideas are cheap and abundant: the key is effective placement of those ideas into situations that lead to action.[5]

You and I have helped create the world we live in, with all its mixed blessings, its paradoxes, its prosperities and its poverties, its loves and its fears, its services and its rip-offs. We have helped shape the way our organizations operate—how we compete and how we collaborate.

To profit, prosper, and be healthy—as organizations and as people—we need to learn how to tap our creativity more, to work with its rhythms, and to become peak performers as defined by our own talents. You *can* make a difference in your organization. And you can make a difference in our world. Don't hold back. As Somerset Maugham put it, "You'll win some. You'll lose some . . . Only mediocre people are always at their best."[6]

CHAPTER 2

You and Your Organization's Climate for Creativity

The Challenges for Creativity and Innovation

▰ *We cannot beat Japan or anybody else by doing exactly what we're doing now, just better. The tack I'm taking is that we have to rethink the car. I'm not talking about making two-wheel gyroscope cars or funny stuff like that, because the customers are not going to change their expectations. I'm talking about the processes and methodologies that would make the car cheaper and faster . . . responding to the marketplace.*

Hulki, chief design engineer for automobiles ◀

The Chinese have a poignant way of communicating "crisis": two symbols are written, the top meaning "danger" and the bottom meaning "opportunity." Crises serve to awaken our inner resources as individuals or organizations, giving us the means to take advantage of the opportunities—*if* we are strong enough to use them that way.

We often don't take action until our environment pushes us to do so; today, our organizations *are* definitely being pushed to be more creative *now*, by a wide variety of forces. The challenge to our organizations is to foster all seven arenas of creativity, putting new ideas into action. This challenge is not a "nice idea," or a "liberal, humanistic dream"— it's a hard, cold necessity. The challenge to all of us is to contribute to our organizations' effectiveness in meeting the following conditions now and in the future.

Q What is your relationship to each of these challenges?

- *The globalization of national economies.* We operate in a "global village" economy. More than ever before, we need rapid anticipation and appreciation of what is happening in world events—the religious and political movements in the Middle East, the buying habits of the Japanese, the international politics of raw materials, even the "import and export of unemployment." Our challenge is to develop innovative practices of economic collaboration for mutual benefit; "them-versus-us" doesn't work within "village" economics.

- *The competitive environment.* After a period of shock from the new success of foreign competition in the 1970s and early 1980s, American business seems to be renewing its drive in technological innovation, updating outmoded management practices, and in some cases making policies to serve *stakeholders* rather than just stockholders. Our challenge is to develop innovative products and production methods and to market with new approaches to an "exchange of value(s)" (perhaps a true definition of "economics").

- *The pace of technological evolution.* Electronics is but one field in which rapid technological advances make products obsolete within a year or two. We live in a world impacted by developments in remote sensing from space; disease-resistant crops; gene splicing; fuel cells; medical lasers, advanced ceramics, alloys, and composites; "smart" membranes; computer-aided design, engineering, and manufacturing; expert systems and artificial intelligence; cellular radio communications; optical computing; nutritional treatment of disease; and so on. Our challenge is to be more than just reactive or even responsive, but to be proactive in *choosing* and valuing what we give birth to technologically.

- *The evolution to an information-based economy.* As Harlan Cleveland points out,[1]

> *A century ago, fewer than 10% of the American labor force were doing information work; now more than 50% may be engaged in it. The actual production, extraction, and growing of things now soaks up less than a quarter of our human resources.*

Information is a new type of resource, more like a flame than an object. With objects, like an apple, if we take some away, eventually nothing is left; with a flame, we can light a thousand candles and they all have flames, including the original. In the same way information is expandable, diffusive, transportable, and shareable. Yet we still often operate with a vocabulary and with management systems based on producing objects, not flames. Our challenge is to learn new ways to foster creativity for flames/information, and new ways to manage and account for the products/services we produce (especially since information may be considered less valuable when it is shared).

- *New and shifting social values/demographics.* Since the 1960s we have been in a period in which seemingly sudden discontinuities with past trends continue to emerge. Examples of these discontinuities include: the 1973 oil embargo; the opening of United States-China relations; the movement to self-responsibility and preventative medicine; the political, social, and environmental protests of the 1960s; and so on. Each shift alters the field of business opportunities, government policy, and social life-style. Our challenge is to develop flexible, innovative plans and responses across a variety of possible future scenarios.

- *Changes in labor force values.* That today's work force has a different mix of personal and work values is not news. Many have cheered the fact; many have bemoaned it. Our challenge is to develop appropriate ways of leading, more than just "managing," the gold-collar (professional), white-collar, and blue-collar work force—and perhaps to challenge the notion that there is any difference.

- *Health, life-style, and stress awareness.* In 1900 only 4 percent of the American population was over sixty-five, and the top four causes of death were the acute, infectious diseases (diphtheria, cholera, smallpox, and typhoid). In 1983, 11 percent of the population was over sixty-five and the top four causes of death were vascular (heart attack, etc.), cancer, diabetes, and cirrhosis of the liver. The most noteworthy aspect of this second list is that, according to the U.S. Centers for Disease Control in Atlanta, 75 percent of the incidence of these diseases is brought about by our life-styles. This self-destructive approach to life is intricately linked with our work climates, our technologies, our attitudes towards stress, and our values. Our challenge is to develop the willingness and ability to use our talents collectively, in harmony with ourselves and with each other; the current lack of harmony *can* be healed for the benefit and profit of all with the renewed exercise of our creative powers.

- *Human survival/prosperity issues.* In 1953, President Dwight D. Eisenhower said,[2]

> *Every gun that is made, every warship launched, every rocket fired signifies, in the final sense, a theft from those who hunger and are not fed, those who are cold and are not clothed. This world in arms is not spending money alone. It is spending the sweat of its laborers, the genius of its scientists, the hopes of its children . . . This is not a way of life at all in any true sense. Under the cloud of threatening war, it is humanity hanging from a cross of iron.*

Our challenge is to develop an economics of value that resonates with our hearts, not just our minds and pocketbooks and to conduct our business planning proactively to bring about the world our hearts desire, rather than "planning" reactively to a future seemingly out of our control.

These are the challenges that move me. They are the ones that give me concern. Global problems are but local problems reproduced too many times. Whether it's in my job or your job, I believe that *what* we produce and *how* we produce it—even in small, entrepreneurial businesses—go a long way to make or break the solutions we need.

Creativity and Innovation Where You Work

So the challenge and the opportunity are to revive and manage our work environment to allow for and encourage creative, responsible, spirited expression. The key question is,

> **Q** How can you apply your creativity to mobilize the creative potential of yourself and the people and resources in your organization?

But before you can answer this question, three other questions need to be addressed first:

1. What *is* innovation and creativity where you work?

2. What are the strengths of the climate for innovation, and what are the areas of improvement?

3. What would you want to improve?

"What is innovation and creativity where you work?" We once asked the brand managers, staff, and division executives of a retail consumer products firm this question. They were in charge of coordinating everything from advertising to distribution for new and existing products, and they wanted assistance in improving their climate for innovation. They pinpointed new ideas for:

- creating products,
- improving current products,
- packaging products,
- reducing production and marketing costs,
- defining new market segments and their needs,

- linking new technology developments to new market needs,

- speeding up the process of pushing good ideas through the system,

- expanding the new-product search process with people inside and outside the company,

- getting customer opinions and developing ways to counteract negatives,

- promoting the company's products,

- stimulating everyone to accept innovation responsibilities and finding ways to reward them,

- looking five to ten years ahead and averting threats to the company,

- expediting logistical plans with efficiency,

- preventing people from feeling too comfortable with the status quo.

Some of the items above relate to creating new **IDEAS,** some to new **MATERIAL** goods, some to **EVENTS,** and some to **ORGANIZING** ongoing groups. Some might involve **SPONTANEOUS** moments. Almost all of them imply being creative in **RELATIONSHIPS.** And the whole project was designed to change the perceptions and **INNER** experiences of each person, to encourage them to be more creative in the day-to-day work world.

> Q What does it mean to be creative and innovative where you work? What would the list look like for your organization? (What would you add or subtract to this list?) What about for your job?

We posed another question to the personnel of the consumer products company: "What are the strengths of the climate for innovation, and what are the areas of improvement?" The participants' responses included:

- Organizational Strengths
  - Innovative ideas are received, discussed, and tested.
  - New organizational structure promotes innovation.
  - Top management is committed to innovation.
  - Management is open, accessible, and receptive to new ideas.
  - Market information is available.
  - Strong financial resources support ideas.
  - Staff has excellent skills and high motivation.
  - Communication has improved within and between departments.
  - People are rewarded and promoted for being innovative.

- Organizational Improvement Areas
  - Develop a more formal process for identifying, testing, and evaluating ideas.
  - Allow more risk in early evaluation of ideas.
  - Improve turnaround time for management approval.
  - Broaden the focus of innovation beyond new-product development.
  - Create a systematic way of rewarding innovation.
  - Reward staff groups as well as individuals.
  - Develop better long-term research to identify future trends and go beyond current capabilities.
  - Communicate management goals and expectations more specifically.
  - Institutionalize management obligations for innovation.

Q What do you think are the strengths and improvement areas for your organization relative to being creative and innovative? What would *you* want to improve?

Compare your organization to the previous lists, or see if you'd like to initiate improvement in any of the following areas:

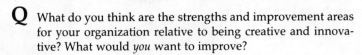

- More training and development on fostering productive work groups

- More encouragement for individuals to be creative product champions

- Better understanding of the marketplace

- More explicit performance-appraisal measures and rewards for creative people

- Better understanding of the needs and strategy for organizational changes

- Broader exposure to methods of being creative, such as brainstorming

- Better understanding of the corporate strategies for growth

- Clearer definition of how new ideas will be evaluated

Fostering More Creativity and Innovation

How can you apply your creativity to mobilize the creative potential of people and resources in your organization? It takes more than just leadership and vision, training people in brainstorming, or installing a new incentive system. There are no prescriptions for fostering creativity, but a systematic approach to fostering innovation is more likely to shift the momentum of the organization. There are eight issues to pay attention to in fostering a CREATIVE climate for innovation:

C = COLLABORATION, COMMUNION, AND COMMUNICATION

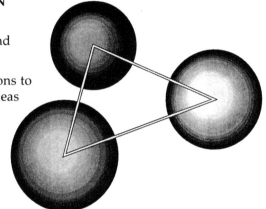

- Collaboration within and between groups
- Development of coalitions to formulate and act on ideas
- Commitment to ideas
- Shared credit

Q What new communication and teamwork abilities does your group need to perform its creative task?

R = ROLES, RISKS, AND REWARDS

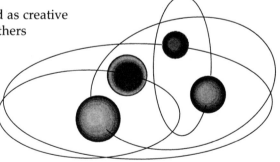

- Individuals perceived as creative by themselves and others
- Individuals allowing themselves and others to take risks (and fail at times)
- Development of appropriate reward menus

Q How does your individual creative-climate process work best? Who are the best initiators of new ideas in your organization, and how do they work best? What rewards do people seek for taking (appropriate) risks?

E = ENVIRONMENTAL MONITORING

- Sensitivity to the environment in which the company is operating

- Establishment of a monitoring system; gathering, evaluating, and disseminating information

> **Q** What trends and events reflect threats and opportunities facing your organization from the outside?

A = ALLEGRO ADMINISTRATION

("to minister to the people quickly")

- Understanding and developing effective systems and processes for actualizing new ideas.

> **Q** In your organization, what are the steps that an idea goes through from its birth in one person's mind to its actualization as a product or service? What hierarchical structures, operating policies, and other factors encourage or extinguish the proper level of risk taking and output?

T = TRANSITION MANAGEMENT

- Comfortable and skillful management of change and innovation within an organization

Q How well is the process of organizational change strategized and managed to promote improved productivity and employee satisfaction?

I = INTUITION AND LOGIC

- Intuition and logic applied to problems to see them from new angles and to challenge assumptions

- Intuition and logic honored in the decision-making process

- Various methods used for both idea-generation and evaluation processes

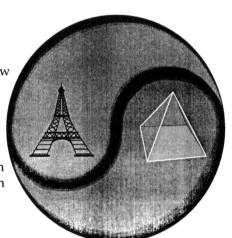

Q How do people consciously provoke new ideas and practical solutions?

V = VISION AND ROLE OF INNOVATION

- A common vision of the company's direction for growth and survival is communicated, understood, and shared

Q What is the purpose of your organization in the long term? What role does innovation play within the organization?

E = EVALUATION PROCESS

- Actively seek feedback
  to learn from successes
  and failures of targeted
  innovations/changes

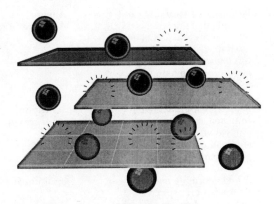

Q How are the best ideas selected so that the most opportune
ideas live but the less opportune don't? When are ideas evaluated and by what criteria?

Notice the types of observations you might make about your organization, which issues they relate to, and whether they might be considered
a strength or weakness:

OBSERVATION	INNOVATION ISSUE	STRENGTH/ WEAKNESS
A group of people struggling with a problem and suddenly coming to a breakthrough, original understanding	Collaboration and Communication	S
A creative and determined person trying to get a new product idea adopted—and breaking the organizational rules if necessary	Roles, Risks, and Rewards	S/W
An individual finding time to incubate a problem, literally "dreaming up" a solution	Roles, Risks, and Rewards	S
A new product failing in the market from lack of insight into consumer values and life-style trends	Environmental Monitoring	W
A person with a new promotion or product idea not knowing where to take it if his/ her supervisor doesn't like the idea	Allegro Administration	W

OBSERVATION	INNOVATION ISSUE	STRENGTH/ WEAKNESS
Performance appraisals and rewards that encourage short-term profits and status quo operations—even when the CEO has stated a need for long-term innovation and a tolerance for mistakes	Allegro Administration	W
People feeling threatened and confused about new, innovative office and factory automation systems being implemented	Transition Management	W
A brainstorming meeting producing uninspired or irrelevant ideas, or ideas that never get acted on	Intuition and Logic	W
Senior management clearly stating a vision of the organization's future and the requirement for market-oriented, innovative new products and services	Vision and Role of Innovation	S
A new idea being killed with questions such as, "Has anyone else tried it yet?" or "What will the ROI be in five years?"	Evaluation Process	W
A new idea *not* being killed, leading to wasted time, money, and effort	Evaluation Process	W

Building a climate with these principles is a creative act in itself and doesn't follow a step-by-step approach. Each of the eight issues overlaps and affects the others. For example, the way that individual roles and rewards are carried out affects collaboration and communication.

The art and science of fostering creativity involve a substantial understanding of where your business is heading and how the current organization can or cannot support actualizing that vision. It also involves the power of your *perseverance*. When a caterpillar is transforming into a butterfly, at first the new butterfly cells are resisted by the caterpillar's immune system. As the new cells become predominant, the immune system switches over and works on the caterpillar cells, assisting rather than resisting the transformation. Organizations, as organisms, often go through a comparable process of resistance, then critical-mass acceptance, and then active promotion of change. This process needs to be honored rather than forced.

Personal power stands behind all of this. Our power is not something we find *as a result* of making changes at work; we are not very effective when we promote change with the assumption that we are starting with no power. Neither is our power related directly to our positions of authority within the hierarchy; CEOs can feel as ineffectual in making changes as the newest entry-level worker.

Rather, realize that our power comes from within us, from our inner sense of purpose, from who we are deep inside. With the experience of our power before we act, our effectiveness will be multiplied.

IN CLOSING . . .

> ◆ *When we were sewing pockets, we were having to recut the pocket pieces plus the pants pieces they were sending us, just to get them to fit together right. That was just losing time and it didn't work worth a damn. We cut a new pattern and sent it off, but they didn't use it for about a year. Then they came back and started cutting the pants differently and used our pockets. Then we didn't have any trouble. So we did make that change.*
>
> *Shirley, seamstress in a garment factory* ◆

> ◆ *When I first started there (at a retail sales company), they had bought a computer that nobody was really putting to use. They were afraid of it, and if you're afraid of a computer, you might as well hang it up. To me it was a challenge. I was young and kind of experimental, so I was the one who took it on.*
>
> *The only programs we had on there at first were the basic payroll and invoicing. I was the only one who was willing to go away for a couple of courses on the machine to actually figure out not necessarily the hardware, but more the software. As a result of what I started producing with that computer, they went to a bigger machine and right now everything is on it including job costing and inventory.*
>
> *Phyllis, office clerical worker* ◆

Creativity is both a work-style and a life-style. The two go together. Life is a twenty-four-hour proposition, and creativity cannot be turned on and off like a lightbulb. Creativity implies full implementation of our potential, not just dreaming up good ideas.

As stated earlier, the opportunity exists for all of us to revive and manage our organizations through the creative, responsible, spirited expression of ourselves at work. What does it take to do this? Be willing to develop the attitudes and skills within yourself to take advantage of the conditions around you and to nurture your own unique talents and experiences as well as those of your compatriots. The eight elements of the CREATIVE climate provide some stimulation and guidelines for doing this.

It would seem logical to organize the rest of this book according to the CREATIVE climate issues: a chapter on Collaboration, another on Roles, the next on Environmental monitoring, and so on. However, there's a more powerful logic to be considered. What makes more sense is to begin with the center of your power to effect creativity and change: you and your own work. From that base, see how you might assist your work group's creativity, and finally your entire organization's. Therefore, Section II, "Individuals Working Creatively," addresses your own competency and creativity in your job. Section III, "Groups Working Creatively," addresses first your work group and then your whole company. Section IV, "Of Profits and Prophets," summarizes key themes of the book and addresses the relationship between technology and human values.

Within this structure the eight CREATIVE climate issues will be addressed, as follows:

SECTION	CHAPTERS	CREATIVE CLIMATE ISSUES
II. Individuals Working Creatively	3, 4, 7	R Roles/risks/rewards
	5, 6	I Intuition/logic
III. Groups Working Creatively	8, 9	C Collaboration/communion/ communication
	10	V Vision, role of innovation
		E Environmental monitoring
	11	A Allegro administration
		E Evaluation process
	12	T Transition management

To close this chapter, imagine, if you will, being in a room with windows on all walls, with shutters closed. Even when the sun comes up, as it inevitably will, no light can get into your room. But if you open the shutters on one window, light can come in. The more shutters you open, the more you can see clearly. It might even happen that you open the shutters at night. Instead of cursing, "It doesn't work," and closing the shutters again, you can patiently wait for the inevitable, enlightening moment.

As this book unfolds, you may discover new windows to open (if you're feeling in the dark), new ways to open them, new attitudes that keep you open to your creativity, and new skills for exercising specific creative muscles. It can give you both food for thought and specific skills for you to develop your creativity at work.

PART II

Individuals Working Creatively

"I didn't actually build it, but it was based on my idea."

The mind is not a vessel to be filled, but a fire to be kindled.

—Plutarch

The starting point for understanding and promoting more creativity and innovation where you work is *you* in your own work, moment by moment and task by task. If you don't know your powers to be creative, much of your potential will lie wasted in an "I'm not allowed to be creative" hopelessness.

Your creativity comes from within you, not from without. You can be creative in your job in private ways. And you can be creative in helping others become more receptive to creative, new ways of doing things.

Some people may believe they *can not be* more creative at work: "If I *were* in the right environment, *then* I could be creative; . . . In *my* company, I'm only allowed to follow rules and policy; . . . Things are so screwed up around here, *no one* could be creative."

Others may believe there *shouldn't be* more creativity at work: "Creativity means people out of control, going off in all sorts of directions. What we need is to put more productivity into how we already do things."

The first set of statements is addressed in this section, and the second set is addressed in part III.

In this section we take you deeper into your potential to be more creative at work, discuss your creative process, give ways to think up more creative ideas, and offer an approach to overcoming personal blocks to being more creative.

In chapters 3 and 4 we offer you a perspective about yourself as a creative person and about your individual process of creativity. In chapters 5 and 6 many techniques for expanding your ability to generate new ideas are covered. In chapter 7 we explore how you can dissolve what you feel are blocks to your creativity.

CHAPTER 3

Developing Yourself as a Creative Individual

Looking into the Mirror: The Creative You

🔊 *When you're talking about (creativity), it's hard to look inside. You know, it doesn't feel creative. I'm really thinking in the sense of being able to go beyond logic—that this is the right thing from an intuitive level. If I believe the house is right for that person, and I'm in there negotiating, I've never lost one.*

In one "impossible" case of no money down and no credit, I just took hold of the challenge, opened a multiple listing book, and started phoning anything that felt like it would fit, asking questions like, "What about a land contract of sale?" which most agents don't even want to deal with. The eventual seller was an out-of-town seller, so we didn't have the benefit of meeting eyeball-to-eyeball. This was another problem to overcome because the financial statement's acceptability is totally dependent on how it is presented. But it worked out.

It goes on the intuitive level. If I really believe that it's the right house for somebody, and if they believe it, then there's a difference in the power of negotiation. If they're not on board, inevitably we don't get the house, even if they say they want it.

—Barbara, salesperson in a real estate office 🔊

🔊 *Few people think more than two or three times a year. I have made an international reputation for myself by thinking once or twice a week.*

—George Bernard Shaw 🔊

🔊 *"There is no use trying," said Alice. "One can't believe impossible things."*

"I dare say you haven't much practice," said the Queen.

"When I was your age, I always did it for half-an-hour a day. Why, sometimes I've believed as many as six impossible things before breakfast."

—Lewis Carroll ◀

Who do you think of as a "creative person"? Albert Einstein? Jonas Salk? Lee Iacocca? Michelangelo? Stevie Wonder? Gandhi? How about yourself? Your manager? Your next-door neighbor?

In our everyday language and experience we may describe a friend or co-worker as being "creative." And we might easily look back in history at the most well-known inventors, scientists, writers, artists, and musicians and label them "creative" . . . at least in comparison with others.

Such comparisons leave the impression that the general population is uncreative and perhaps unable to be creative. Rumors abound that "Once you're past a certain age, you're no longer as creative." Who, or what, is the "creative person"?

The context in which we view our creative abilities is critically important. It is *much* easier to refocus our creative abilities onto new, different realms than to think we must go from being uncreative to creative. This chapter and the next can help you enhance your CREATIVE climate for innovation by focusing on individual "Roles, Risks, and Rewards" in being creative.

It may seem that some people are "naturally creative." Willis Harman, a regent of the University of California, says:[1]

> Some (people) seem to have stumbled into the extraordinary moments accidentally, with no conscious intention of seeking them, while others seem to have learned to consciously "invoke the muse" by following certain steps that triggered the breakthrough.
>
> There seems to be something in all these accounts (of creative insights), despite their sharp individual differences, that speaks of a capacity for creative breakthrough—a capacity that might be independent of talent, or field of endeavor, or life circumstances.

You are a creative person, perhaps in many ways you don't recognize. There is an entire spectrum of creative imagination ranging from the mundane to the miraculous, a spectrum greater than most of us would ever imagine.

You are always being creative. In chapter 1 we saw that everyone can exercise creativity in up to seven dimensions: "idea," "material," "spontaneous," "event," "organization," "relationship," and "inner." It is *what* we spend our time creating—not *whether*—that makes the difference between calling ourselves creative or not.

You are naturally creative, and so am I. Our creativity springs from our unique differences as much as from the human nature we share. Creativity begins with the recognition of our differences. By discovering our uniqueness, we can know our own will as an integral part of the universe. Awareness of this will set in motion the act of creation to express our uniqueness.

Your creative self-expression is promoted by a balance between your inner peace (which allows you to hear your inner wisdom) and positive external action. A primary aspect in our ability to create the life we want is commitment, the *intention* to have our lives reflect on the outside what we will and feel and think and dream on the inside.

I'm sure you've noticed how some people accomplish what they set out to do, whereas others make up all sorts of "New Year's resolutions" and never achieve any of them. The lack of intention is just one way that we restrict our ability to express ourselves creatively. Ironically, some of the ways we stop our creativity—whether from too much comfort with the known or too much fear—are truly ingenious!

However, it is more important to focus on the positive habits for using our creative potential than to concentrate on some "blocking" aspect of ourselves. Although understanding how we stop ourselves from being creative can be important—and in chapter 7 we explore ways to go beyond these stopping points—we need to focus on more creative ways of being and let inhibiting habits wither from disuse.

What's a Creative Person?

Creativity comes in as many forms as there are people, yet do the people we label "creative" share some common habits? Is there a model of the creative person within which you might see your own potential in a new light?

Take, for example, Einstein. It may seem like a lofty ambition to emulate his creativity, yet he was very human; besides his "professional" work in physics, his two favorite creative pastimes were daydreaming—especially beside water—and playing the violin.

Henri Poincaré, the great French mathematician, theorist, and philosopher of science, once said of Einstein:[2]

> Einstein . . . does not remain attached to classical principles, and when presented with a problem in physics he quickly envisages all its possibilities. This leads immediately in his mind to the prediction of new phenomena which may one day be verified by experiment.
>
> I do not mean to say that all these predictions will meet the test of experiment when such tests become possible. Since he seeks in all directions, one must, on the contrary, expect the majority of the paths on which he embarks to be

blind alleys. But one must hope at the same time that one of these directions he has indicated may be the right one, and that is enough.

Although it is always dangerous to stylize or generalize about traits by trying to identify what creative people are like, such a discussion can help show us ways to exercise more power and judgment in how we are creative. When we try to characterize the creative person, we often refer to the "inventive" qualities. For example, one creativity researcher says that "The creative individual not only respects the irrational in himself, but courts the most promising source of novelty in his own thought."[3]

But this is not enough. More than inventiveness is involved. To develop myself and my work habits, I've long been interested in understanding "original thinkers," great artists, and pioneers in business. There are two sources I draw upon most: the works of Abraham Maslow, "father" of the humanistic psychologies in this country; and of Dr. Joseph McPherson, former manager of the Innovation Program at SRI International.

Abraham Maslow developed a new notion in Western psychology when he mused that our ideas of the optimum person have mainly come from observational study of the mentally disturbed (through Freud) or scientific study of animals (through Watson and the behaviorists). But that's like visiting the neighborhood sandlot to discover the best athletes. Instead, he metaphorically went to the Olympics and tried to discover the commonalities among the Einsteins, Michaelangelos, and Lincolns. These individuals fully expressed themselves and their talents. Maslow coined the term "self-actualized" to describe such individuals.

Dr. Joe McPherson has observed people in the workplace whom others have called creative. He has identified certain commonalities among them, habits that can support our inherent creativity. These commonalities can be grouped into five categories:

1. Self-orientation

2. Motivational characteristics

3. Attitude toward others

4. Attitude toward past, present, and future

5. Intellectual characteristics

What's important for this chapter is the "bottom line" of the work of Maslow and McPherson. Their work can be combined to give a model that can sharpen our view of our actual and potential creativity.

As you read through the model given below, identify what you see in yourself and what you see in others you call creative. Remember, what you first recognize in others whom you call creative will often be the next thing to awaken in yourself. That awakening will occur as a ripple effect—by concentrating on one or two parts of the model, other parts will grow in you without effort.

A person who fully expresses his or her creative potential is a SPIRITED person. Your creativity is based on key strengths in a *few* of the following characteristics. As a creative person, you do not have to be everything listed in the following boxes. If you recognize your strengths and emphasize your own way of being creative, you will recognize yourself as truly SPIRITED:

S = SPONTANEOUS

- Fresh
- Childlike, at times
- Curious
- Willing to take risks
- Sense of humor

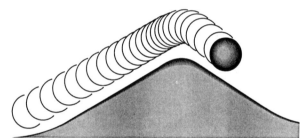

P = PERSISTENT

- Energetic
- Courageous
- Assertive
- Independent
- Determined

I = INVENTIVE

- Looks at problems in new ways
- Likes challenge
- Sometimes skeptical
- Comfortable with ambiguity

R = REWARDING

- Willing to share credit
- Values personal satisfaction, peer recognition over money

I = INNER OPENNESS

- Intuitive
- Can switch from logic to fantasy easily
- Open to emotions
- Promotes peace and love
- Can think and act/create and innovate in different modes.

T = TRANSCENDENT

- Can see situations realistically
- Fantasizes how he/she wants things to be
- Confident can effect change
- Chooses growth over fear

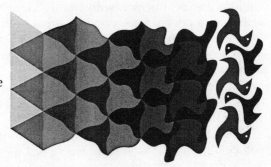

E = EVALUATIVE

- Discerning, discriminating
- Judgmental at appropriate times

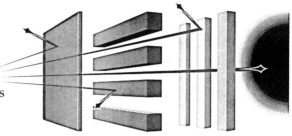

D = DEMOCRATIC

- Values and respects people
- Identifies with mankind
- Seeks stimulation from variety of people
- Responsible
- Promotes highest benefits of *all* concerned

Do you recognize these characteristics in those you would call creative, including yourself? I personally find the spontaneous and philosophical responses hard to muster sometimes, while the inventive and democratic ones come much easier. The energetic and intuitive characteristics seem to come and go. However, if these potentialities weren't already within you and me, we wouldn't recognize them at all.

It is quite evident that the key characteristics of the creative, self-actualizing person are courage and persistence—the courage to risk, to (ad)venture into new territory, and to be a minority of one with a new idea. An internal IBM study of eighteen projects concluded that the major difference between success and failure in the projects was the presence of (or lack of) a determined, persistent innovator ("product champion").

And Nolan Bushnell, founder of Atari and Axlon Toys, has said he would rather be a person with twenty ideas, of which five might work well, than a person with two ideas that both work well. That way, he figures, he's three ideas ahead of the other person. "Unless you're willing to accept failure," Bushnell suggests, "you're not really willing to push yourself to the edge."[4]

The other SPIRITED characteristics can be equally important to consider. All of them are identifiable simply because they are repeated patterns of thinking, feeling, and behaving.

You are a creative, adventurous, peaceful, loving, productive human being. The characteristics we've just discussed are like items on a menu. We can pick and choose those that represent our strengths, and represent a way of being we would like to actualize in ourselves. We do not have to be everything at once. We will exercise different aspects of ourselves at different times and in different ways.

Like Maslow and McPherson, we each need to make the same investigation of people in the everyday work world, observe ourselves and others for these characteristics, and discover the optimum practice of our inherent creativity!

Polishing the Mirror

If we are all naturally creative, why does it seem that only a few people exercise their creativity? Primarily, we may underdevelop, or we may "misplace," personal habits for being creative—often from fear of "bucking our culture" if that culture doesn't support our version of creativity. Erich Fromm states that the creative attitude first requires the capacity to be puzzled. While most children still have this capacity, most adults believe they ought to know everything—that to be surprised or puzzled by anything implies ignorance.

Perceiving the characteristics of the SPIRITED person is the starting point. But the list by itself only gives us interesting information that is likely to be half-forgotten as soon as the book is closed. The important issue is how we see ourselves in the present and the future as creatively expressive individuals.

At most advertising agencies there is a "creative" department. Imagine a research scientist at a cocktail party asking someone from the advertising agency, "What do you do?" and getting the reply, "I 'do' creativity." The scientist could easily answer, "I do creativity also." As you might guess, we could all answer the same way about whatever work we're in.

Yet, creativity is fundamentally an expression of *who we are*, not what we do or have. The true force of creativity is first to "be" (for example, seeing yourself as a doctor), then to "do" (going to medical school, conducting examinations), and finally to "have" (a diploma and a practice). In the natural scheme of creativity "being" comes before "doing," which comes before "having."

Unfortunately, many people try to "have" (the diploma) so they can "do" (conduct examinations) so they can "be" (a doctor). All along, these people are trying to create from a sense of *not* being something, and it's therefore hard to draw a positive mental picture of finally being, doing, and having what they want.

Start with a vision of who you are and who you want to be, rather than what you want to do or have. Then build in the "doing" and "having" to fill out the vision. A focus on doing or having simply doesn't pack the power or follow-through. "Losing weight" or "giving up smoking" or "having a Mercedes Benz" usually becomes a dead resolution—and sometimes, a source of guilty feelings—unless such a goal is directly consistent with a desire from deep inside to "be" all you can be.

Take the time to imagine yourself as the creative person you want to be. This is an important step in your developing a creative edge—an internal edge as well as an advantage in your work. Your imagination is a tremendously powerful tool in bringing about the skills and attitudes to support your creativity.

By analogy, consider the example of a famous study of visualization as a method of improving basketball players' free-throw shooting. Three groups were tested and judged to be equal in their abilities to shoot free throws, as measured by the percentage of successful free throws. Then one group was allowed to practice more; the second group was allowed to stand at the free-throw line and imagine making successful free throws; the third group was given time off and didn't appear on the court. When the three groups were tested again, the group that had actually practiced with the basketball had improved, as you might expect. However, the group that had only visualized shooting improved almost as much. The third group showed no improvement.

Athletes have shown us much about the power of mental rehearsal under "relaxed attention" states. The method can affect even our unconscious body movements. In the same way, rehearsal with models of ourselves as creative can help us awaken the expression of our creativity more and more.

To help you see in yourself the characteristics of creative people, a personal exercise may prove quite valuable to you. Read through it first, and then please *take the time* to close the book and actually do the exercise. The insights that occur to you during the first reading are important, of course, but there is more to find in the exercise to get its full benefit.

Let things happen as you would wish them to. Don't let your internal idea killers come out and inhibit your creative dreaming with, "That could never happen," or "He (or she) would never go for that." Create the vision, then update it later. That way you let yourself stretch and grow while still staying "realistic."

With that said, please read on.

> On a separate piece of paper, write down two or three moments in which you observed someone else in his or her moments of creative expression. These should be moments you've witnessed, as much as possible, rather than just recalling that, say, Michelangelo had some inspiring moments painting the Sistine Chapel. The creative moments can involve ideas, material/sensory creations, spontaneous moments, events, organizing, relationships, or inner experiences. How was this person SPIRITED?

Now write down two or three more moments in your *own* life when you experienced the more inspired, creative moments . . . not just in coming up with innovative insights, but actually carrying them through and fulfilling them. How were you SPIRITED?

Pick a current situation in your life in which you would like to be more creative or in which you want to find a creative solution to a problem. For a moment, imagine yourself to be one of the other creative people you described in the first part of this exercise; close your eyes and really pretend you are this other person. Vividly imagine this person's feelings and beliefs about himself or herself. Feel the motivation, think the thoughts, act the actions of that person. Be SPIRITED.

As this person, imagine being creative in the current situation you described. How is the person—or how are you as the person—being creative this time? Carry the fantasy through as far as you can, even if you don't come up with a solution you like.

Now, close your eyes again and recall how you felt, thought, and acted when being personally creative. Really bring moments alive for yourself. Now, while living that creative experience, imagine being creative in the current situation you described. How are you going about being creative this time? Be SPIRITED.

Carry the fantasy through as far as you can.

Now that you've completed the exercise, please try it once again. This time close your eyes and follow the directions using your powers of imagination and intuition.

The more carefully we focus on who we are deep inside and what we want to create, the more likely we are to act creatively at work. We become that which we dwell on, that which we give our attention to. Our thoughts and our words become real. That is our creative power.

If we dwell on the very things we fear or don't want, such as negative habits, we help bring them into our lives more. But by focusing our attention on ourselves in our most creative moments—or by "borrowing" the creative experiences of others to try on for a while—we can evoke and bring forth the optimum of our creativity in any situation, with any problem.

What's in It for You—Or Others?

We live and work by relating to others, not by being in a vacuum. Our growth in creativity occurs most rapidly when we set up our environment to support us. Like a rose bush in nature, we need our water, air, soil, and fertilizer.

The key to this environment is setting up the feedback, recognition and rewards, and interpersonal support that we most thrive on. While in some cases this may mean changing jobs, there is more we can do than we may realize to "manage our bosses" and our workplaces.

The Feedback

Feedback is the answer to the question, How do we know when we're successful? Sometimes we are afraid of failing, so we refuse to open channels of feedback. In doing so we not only keep ourselves from knowing about failure but also from experiencing our successes. Only with good feedback channels can we experience personal satisfaction and a sense of accomplishment.

For innovation to thrive the feedback must be tolerant of mistakes. For example, an ad for 3M headlines: "3M has made a lot of mistakes. We're very proud of some of them." It goes on to say, "Everyone who is alive and moving makes mistakes. The trick is to learn from your mistakes and move on. [If venture risks are reasonable], we tend to be willing to make an investment and learn."

Recognition and Rewards

We as people have very different values and very different "vitalizers" to keep us interested in our work. Some prefer outer signs and symbols of success, whereas others prefer internal, self-satisfying experiences. The question of motivation and benefit is a question of what we value and of being empowered to be what we can be.

On the surface, empowering people and satisfying their values may seem an administrative nightmare, yet it is precisely what the IBMs of the world attempt to do. At IBM there is an award system to recognize both technical and nontechnical individual achievements. This is one way of showing employees that achievement is important and that they *can* control their own destinies.[5] Top management gets involved in the recognition, and the award winners are publicized widely. Although IBM typically awards money, there are other forms of recognition:

- Attendance at conferences

- Membership in professional societies

- Service on local organization committees

- Publicity in organization newspapers

In addition, managers reward their staff in less tangible ways by:

- recognizing the need for relief from the routine and providing new challenges,

- asking for advice on a problem,

- showing in the presence of others that you understand what an employee is doing,

- helping employees add to their skills,

- publicly defending the work group,

- seeing to it that top management recognizes an employee's contribution,

- leaving employees alone for a while.

The important thing is that the benefit fits. For example, when a creative person is put in charge of an internal new venture, an appropriate alternative to offering financial incentives for success is to offer control of larger venture budgets with the freedom to tackle a challenge that is personally inspiring.

Kodak's highly successful Office of Innovation has helped to bring many people with new ideas together with financial sponsors who have no reporting relationship with the idea people. The program has grown over the last six years from one office in one laboratory to about twenty offices worldwide, yet there have been no financial incentives to induce people to come forth with new ideas. The key is giving potential entrepreneurs the possibility of becoming the general manager (or at least a key manager) of a new business.

Whatever your position in the hierarchy, you can develop appropriate reward menus for yourself and others. The goal is an environment flexible enough to foster peak performance for each individual, not just a few key people and not just the common denominator of the average employee.

At Mervyn's, chairman Jack Kilmartin and all other executives send out note cards that have "I heard something good about you" printed at the top to clerks, buyers, trainers, and other line employees, as well as managers. GE Medical Systems uses "Attaboy" recognitions (a written acknowledgement or pat on the back) even for seemingly minor accomplishments. We can follow these examples anytime, as a receiving clerk, a computer programmer, a manager, or an executive.

Support Networks

One of the main reasons some people work creatively is to gain the respect of their professional peers. Giving and receiving this respect highlights one of eight ways in which we all need to give and receive personal support. The eight are:

1. Listening—really hearing and mirroring what someone is saying without offering advice, judgment, or other personal input.

2. Unconditional positive respect—showing respect and/or love (even when disapproving of someone's behavior)

3. Emotional appreciation—reinforcing someone's right to his or her emotions and viewpoints

4. Emotional challenge—not letting someone limit herself or himself by personal fears

5. Talent appreciation—acknowledging work well done

6. Talent challenge—showing someone his or her mistakes or next steps for growth without personal put-downs

7. Touch—having appropriate physical contact, including handshakes and hugs

8. Play—being able to have fun and laugh with or without a work focus

Q At work, from whom do you receive each of these types of support?

Are any of these types of support easier for you to allow yourself to receive?

If you feel a deficiency in any area, from whom could you get the specific support if you asked for it?

Can you develop peer support groups within or outside your organization?

At work, to whom do you give each of these types of support?

Are any of these types of support easier for you to give?

Who do you think/feel could use a specific type of support from you?

You can deliberately develop support in your life. You can discuss your work goals and set up each type of support to help you fulfill your responsibilities as creatively and effectively as possible. This is a key to giving yourself and others the right blend of challenge and appreciation—and to developing yourself as a more SPIRITED person in a more CREATIVE climate for innovation.

IN CLOSING . . .

■ *Top management implemented a cost-containment program where if any employee, other than managers, could find a way to save the hospital money, they would get 2 percent of the first annual savings, up to a limit of $3,000. So people like housekeepers and other people were all of a sudden putting their minds to work, on how they could save the institution money.*

They're still using the program and they're not only paying out cash, they're finding a lot of rewards financially because people are actually thinking of how they can do their job better—like in the cafeteria and how we order things.

Trena, manager in a hospital cardiac surgery unit ◄

■ *I think you'll find that most senior scientists are driven by a lot of inner drives. They are naturally curious people who wonder why. In particular, if you ask them what they really want out of life, they'd like the respect of their peer groups. They want to be able to give a talk on what they've done and have people say, "Not only did he do good work, but look how clever he is. Look how he thought that out. Look at what an elegant thing he did. It wasn't just 'good,' it was a very elegant approach."*

You could be a good skilled worker, let's say, and do everything correctly and well (but) we're judged by whether we've done it elegantly, nicely, and cleverly—They're judging the creativity, not the tools.

Bob, university professor ◄

There is a peculiar source of tension that arises only from inside us. It is the tension of not exercising our talents, not expressing our creativity and aliveness. We can become numb to this, like the apathetic workers who have already retired mentally but who still have ten or thirty years to go before qualifying for retirement benefits. This tension for growth can only be released by awakening our creative expression, at first perhaps at work and then elsewhere in life.

The source of your creativity is in you, not in your environment. Even the inspiration to set up your environment to support your growth comes from within. By analogy, Kabir, the fifteenth-century Persian poet, says:[6]

> The jewel is lost in the mud, and all are seeking for it;
> Some look for it in the east, and some in the west; some in
>   the water and some amongst stones.
> But the servant Kabir has appraised it at its true value,
>   and has wrapped it with care in the end of the mantle of
>   his heart.

CHAPTER 4

Enhancing Your Individual Creative Process

Setting Your Sights

▲ *Is a part of creativity having the ability to bring the pieces from far apart and visualize your formulation in the mind—is that important? In my current work I'm taking from four fields: polymer chemistry, carbon fiber physics, catalysis chemistry, and silicon nitrogen chemistry. I found that other people had finally made polymers for making silicon carbide and nitride, (but) the compounds were really inefficient. I knew that if I could activate silicon connection bonds, there might be something interesting. So I looked for a way to make the silicon nitrogen polymer using a catalyst that breaks silicon nitrogen bonds.*

I realized (after further developments) that it was a general way to make just about any material that is otherwise made at very high temperatures, because the way you form things at high temperatures is very difficult.

Rick, lab director in organo-metallic chemistry ▼

In September 1973 Watergate was in the news, and the oil crisis was yet to strike. That month I began coteaching a college course called "The Year 2000" with now-Congressman Newt Gingrich. The first exercise we gave to the students—and ourselves, since we had never done it—was to write an essay describing a plausible picture of our own lives and life in general twenty-seven years in the future. Then we each wrote an "autobiography" of how our life progressed up to the year 2000.

Thus, we jumped ahead to create a vision of ourselves and our world and then looked back to detail how that vision might come about. This is a wonderful technique for making what we want of our lives. I, for example, already felt strongly about becoming an international business consultant, though I didn't anticipate the winding path it took for me to become one.

When we went around the class and had each person read his or her "autobiography," a chilling realization dawned on us. At least half of the stories had the theme, "It's the year 2000 and the world is falling apart. I'm living on my twelve-acre farm in northern Canada, and I just hope I die before it falls apart up here too."

We explored how our present assumptions about the future have a great effect on the energy and determination we put into living our present lives. We saw quite clearly that self-fulfilling prophecies *do* happen. With strongly held assumptions about gloom and doom, we would be very apathetic to the notion of social and work responsibility. The chances would be increased for the doomful future to come about in ideas, material forms, spontaneous happenings, events, organization, relationships, and inner experiences.

Furthermore, we realized that without any vision at all, we operate reactively in the world, letting events outside of us dictate our alternatives. With a vision we operate proactively, shaping our world in line with our positive or negative vision of the future.

Is there a "formula" for this creative shaping?

It's Your Process

Creativity is not something we can turn on and off like a faucet. It is an experience and expression in our lives that must be nurtured. Yet the moments of creativity seem to come on their own. This nurturing process means that creativity is at once a skill, an art, and a life-style. There are specific steps we can each take to bring out our creativity in our own unique way.

Each of us has a unique style and rhythm in being creative and innovative. However, there are similarities in the types of activities it takes to create. To help you understand your way, this chapter explores a way of understanding the general process of creativity—applicable to developing ideas, material form, spontaneous moments, events, organization, relationships, and inner experiences in the work situation. Later chapters explore the creative process in groups.

Instead of taking it as "gospel" on *the* individual creative process, use it to gain more clarity on how you can be more creative. See what aspects you are most comfortable with and which you want to develop better. See what steps you take, and in what order, when you are most creative. See how you can more deliberately bring about the conditions when the muse visits.

The process of idea generation can involve important periods of incubation and inspiration. Even beyond that, the individual creative process may be seen as having seven basic activities. These comprise the APPEARE process by which things "appear" or get created.

A = Be AWARE of your complete current SITUATION

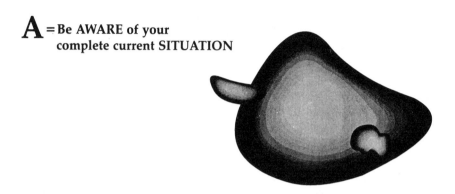

P = Be PERSISTENT in your VISION

P = PERCEIVE all your ALTERNATIVES

E = ENTERTAIN your
INTUITIVE GUIDANCE

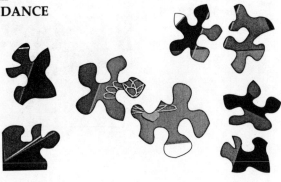

A = ASSESS and select
among your ALTERNATIVES

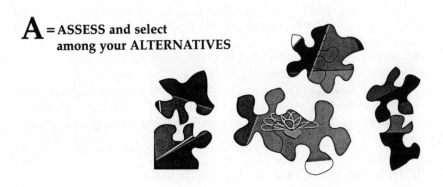

R = Be REALISTIC
in your ACTIONS

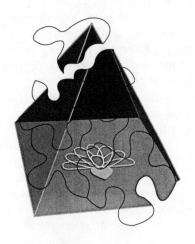

E = EVALUATE
your RESULTS

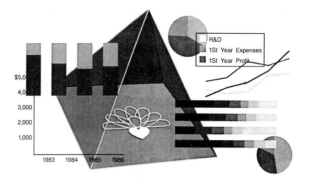

The activities may not always follow in the same sequence. Sometimes you may not have a clear vision until after perceiving and assessing your alternatives. Sometimes you may know intuitively what to do without any conscious consideration of alternatives.

Persistent vision, perceived alternatives, and intuitive guidance form a minicycle in this process, as noted earlier. For example, the vision may first become clear, and then the alternatives and intuitive guidance will show a particular solution. Or the intuitive guidance may first give alternatives to explore, allowing the vision to become clearer.

But it is useful to have a model of the creative process to make sure no activities are left out of your individual method. It may take many times through the entire set before an "Aha!" is reached and action is taken.

Be Aware of Your Complete Current Situation

Making a scientific discovery may depend on full analysis of all available facts and research. Creating a successful meeting can be based on knowing the people who will attend, the decisions to be made, the structure needed for the agenda and meeting process, and so forth. Painting a picture requires selecting pigments with the right features and colors and instruments (brushes) with the right suppleness.

Although these awarenesses may or may not come at the very beginning of the creative process, they represent "grounding" the process in reality. Writer Thomas Wolfe once stated that this awareness or inspiration rarely comes unless we have immersed ourselves in a particular subject. This grounding is critical to problem solving in business situations.

In the same way, companies that are very aware of the complete current situation of their customers' suggestions reap great benefits. Eric Von Hippel at MIT conducted a study of scientific instrument and component equipment manufacturing businesses. In a review of 160 inventions, over 70 percent of the product ideas came from users. And it wasn't just the special features that came from users. They inspired over

60 percent of the minor modifications and 75 percent of the major modifications. And surprisingly, *100 percent* of the so-called first-of-type ideas (complex instruments such as the transmission electron microscope) were from users.

Complete awareness of the current situation means exhaustively looking for all available information. Useful information may come from "unexpected" sources, both internal and external.

Be Persistent in Your Vision

After Einstein had described some previously mysterious orbit behavior of Mercury using his new gravitational field equations, he said,[1]

> *In the light of knowledge attained, the happy achievement seems almost a matter of course, and any intelligent student can grasp it without too much trouble. But the years of anxious searching in the dark, with their intense longing, their alternations of confidence and exhaustion, and the final emergence into the light—only those who have themselves experienced it can understand that . . . I was beside myself with ecstasy for days.*

A vision or goal is meant to guide our efforts. With consistent energy in one direction we have the best chance of getting there easily and efficiently.

Actually, vision itself is guided by a higher sense of purpose. Your purpose expresses your integrity and what you stand for consistently over time. It is your sense of fundamental values and meaning. It is your deepest "being true to yourself." Your visions of your future represent specific ways you might express your fundamental purpose.

Look back on the various times in your life, from childhood on, when you felt most alive, joyful, fulfilled, "purposeful," "in touch," or creative.

Q   What do those moments have in common?

What pattern can you see that reflects your deepest sense of mission in life? To live here and now and experience what comes along? To help bring about world peace? To develop a large, profitable business? To raise a family that lives the Ten Commandments daily?

What do they tell you about yourself?

What do you value?

Ideally, why do you work?

Your creativity is always related to bringing forth something you value. Expressing your creativity *more* does not imply that you must change your core values. It does imply that you improve the quality of your expression of purpose and values. If you value safety and security to the point that you don't take any of the risks needed to be creative, you will need to redefine what safety and security mean to your core purpose. With your purpose clearly in mind you can most easily create the right visions, circumstances, relationships, and other resources to support your creative energy.

Creative moments are nurtured by holding a vision with persistence. The writer with a vision of a completed book is guided to work persistently six to ten hours a day, giving a chance for the muse to visit. The entrepreneur with a persistent vision of a thriving business keeps his/her energy directed for the one to three years it often takes to get a business on secure ground. A manager with a persistent vision of a work climate that's more open, trusting, and productive employs a patient strategy for promoting change.

Maintaining a persistent vision really requires five actions:

1. Make the vision specific.

2. Create a clear idea or picture. Imagine it vividly in the present tense, as having already arrived. Use all your senses—sight, taste, touch, smell, and hearing—in your imagination.

3. Focus on it often. Bring the vision to mind in daydreams, meditation, or whenever you can concentrate.

4. Give it a positive energy. Make strong affirmative statements to yourself, stated in the present tense, seeing yourself in the new state.

5. Acknowledge doubts and disbeliefs without investing energy in holding onto them.

Sometimes this vision is clarified as a result of awareness and analysis of a situation. Sometimes it seems to come more spontaneously. In all cases, the deeper the vision is felt, the more likely you are to pour your energy into making it a reality.

St. John's Gospel begins, "In the beginning was the Word . . . and the Word was made Flesh." Your vision is your participation in "speaking the word" so that it can be made real, for the highest good of all concerned.

Perceive All of Your Alternatives

If you're like me, when you face particularly troublesome situations, you tend to rush to find a solution. In all likelihood, many plausible, alternative solutions are never considered. If I only had a nickel for every time I've had to backtrack. Now I try to remember an old African saying regarding crossing dangerous or wide rivers: "Sometimes the shortest way there is the longest way around." "Hurry, go slowly," is another way of saying it.

One of the most stimulating—even humorous—parts of the creative process is dreaming up alternate ways to approach the situation. Artists do sketches. Thinkers brainstorm. Writers, even writers of letters and memos, make different outlines and try different wordings. In chapters 5 and 6 I describe many avenues for you to follow to generate new alternatives, especially when you feel "stuck" and nothing is coming to mind.

Be aware, however, of how we often stop ourselves from identifying our creative options with *idea killers*. Idea killers are the thoughts we think or hear from others that awaken our fears of being impractical, stupid, unsuccessful, and on and on. These idea killers are the statements we say to ourselves or others, like:

- "Oh, that will never work."
- "That's a crazy (or dumb) idea."
- "Top management will never go for it."
- "That's already been tried."
- "That idea is ahead of its time." (Sounds like a compliment, doesn't it?)
- ——— (silence)

When generating alternative solutions (insights or courses of action), it is very important to focus them within the scope of the problem, vision, or goal yet refrain from using habitual idea killers. We must eventually evaluate the alternatives, of course. However, the creative process works best when we separate the evaluation step from the alternative-generation step. Keeping them separate not only opens us to generate alternatives more freely, it also helps us to avoid rushing to a solution by judging too hastily, "That's the perfect solution!"

Entertain Your Intuitive Guidance

The inspired moment of invention or composition is often experienced as the muse visiting. In that experience, it seems as if we are just the conduit through which the creative product is being born.

For example, Melvin Calvin, honored as a Nobel Laureate in Chemistry (1961) for his work in photosynthesis, reports that he had been waiting in his parked car when he was struck by an intuitive flash. In a matter of seconds the cyclic character of the path of carbon became apparent to him—the recognition of phosphoglyceric acid—all in a matter of thirty seconds.

In such cases, the "activity" for us to follow is to surrender and allow our intuition to give the answers we seek. The importance of the intuitive part of the creative process cannot be overemphasized. Sometimes we may want to control or "will" a creative answer, and overachieving efforts can block the intuitive process. Relaxation plays a key role in letting our intuition work and communicate with us.

This is not to deny the role of persistence or logical analysis in creation. But the hard work of creativity must include some relaxed, intuitive time for assimilating information and envisioning solutions.

Our inner, intuitive self compiles both verbal and nonverbal information, both logical and sensory/emotional input. From there, the muse takes over. Fluency in both right-brain imagery and left-brain languages is an invaluable aid to creativity. Since the language of intuition is sensory (images, sounds, etc.,) instead of words, we must pay attention to the symbols and translate them.

Einstein labored over logical data for years, eventually "saw" intuitively his theory of relativity, and then translated his insights into words and mathematical formulae. Both the insight and the translation were creative. Only when he could communicate it to others were the full creation and the full value of his theory complete.

Logic has often been considered a "masculine" attribute and intuition a "feminine" one, corresponding to the left brain and the right brain. We all have both capabilities, though perhaps in differing degrees, just as we all have left and right sides of our brains. To be creative and innovative we need a fluency in masculine and feminine qualities in ourselves. Likewise, as our organizations become more fluent in masculine and feminine creativity, we will have more "edge" in our creative climates.

Finally, it is common to confuse the voice of our inner, intuitive self with that of our emotions. They are not the same, though the emotions might provide information for intuitive musings.

With practice in inner quieting, you can perceive the subtle but important difference between following your intuition and following your emotions. Your emotions are more often tied to your perception of the world around you and your judgments about that world. In contrast, your intuition is your perception of an inner world that is clear and wise. When the voice of your intuition is not drowned by the noise of everyday thoughts and emotions, you can recognize its clarity and wisdom.

Assess Your Alternatives

There are many ways to assess your alternatives, including the following:

- *Criteria Checkerboards.* Each alternative is listed on one axis of a matrix, and each criterion is listed on the other. Each alternative is rated on a specified scale (1–10, high/low/medium, +/0/−, etc.) for each criterion, and the checkerboard, shows you the relative merits of the alternatives.

- *Rank Ordering.* For each criterion, each alternative is compared to each other one. The "winner" each time gets a point. Then the scores are totalled.

- *Sorting by Category.* Alternatives are grouped into categories and evaluated as synergistic "clumps" of solutions, according to a criteria checkerboard, rank ordering, or other means.

There are also many ways to select then among your alternatives, including the following:

- Choosing combinations of solutions

- Building a solution starting with one acceptable component

- Eliminating unacceptable alternatives until an acceptable solution is left

- Backing off and *not* deciding

Whatever the methods, there are two keys to successful assessment and selection: being open to the *best solution* for all those affected by the decision (the stakeholders); and using our *intuition and emotions* in conjunction with our *analytical* abilities.

The Best Solution　"Letting go" of our ego or will and surrendering to a higher will can ultimately bring us more satisfying and productive decision making. The solutions tend to be more easily and effectively implemented.

　　Sometimes we might have a pet solution, a hidden agenda, an important self-interest (financial or otherwise), or a desire for a "convenient" solution. Or we might take persistence to the point of narrowly focused stubbornness. These can hinder us in objectively seeing the highest good of all.

Using Intuition and Emotions　At times facts and figures can be manipulated to support all sorts of ideas, sometimes at odds with what seems intuitively correct. Whether dealing with a new customer or a new idea, getting a feel for the situation and acting with that extra "data" can make or break a decision.

　　At other times organizations have tried to appoint a product champion for an idea that was attractive on paper ("in theory"). They have discovered that if no one is excited about a market opportunity or a new product idea, it is probably doomed to mediocrity or total failure. With enthusiasm, people will work well beyond what's formally required to see "their idea" become a proven innovation.

Emotional enthusiasm and "gut feel" play a vital role in assessing and selecting among alternatives. As one practitioner in the health industry states, "I choose my next step by my goose pimples. If they're there, I know I'm on the right path. If they're not, that step is probably not right for me."

Be Realistic in Your Actions

To bring to life an invention, a creative marketing strategy, a report, a piece of art, or a better working relationship, you must *act*. Although both logic and intuition play key roles in decision making and the creative process, the creation is usually stillborn without realistic action. Just because we have suggested an idea doesn't mean our work is done. It usually can't be left to someone else to work out the details and implement our idea.

Creation requires action, but action is not the same as activity. Action is based on inner awareness; it targets and achieves only that which is harmonious with our best interests. Activity pursues habits that may not serve us anymore. For example, with action we eat when we are hungry. With activity we eat when the clock says we ought be hungry, whether we really are or not.

Action often takes courage. A new product idea must be documented, supported with information, discussed, and even fought for in order to flourish. An Einstein must write his formulae *and* defend his data and insights.

Realistic action is based on freedom to act, which in turn is based on commitment. Commitment in action means unfolding our potential to become bigger than the situation. Without commitment, we remain in a state of analysis-paralysis, stalemating our talent as an eternal apprentice in a world of masters.

Commitment is more often discovered within ourselves than made up. It's where we find our goose pimples. And commitment frees us, rather than restricts. Dancers with commitment learn skills that allow them to express movements most of us cannot; they have developed and earned a freedom most of us do not enjoy.

Commitment, however, is different from attachment. It means acting with the best knowledge available, then emotionally surrendering to whatever results, then learning, then acting again with the best knowledge available, then emotionally surrendering to the results, then learning, and so on.

This is the cycle of staying in a problem-solving mode that doesn't sacrifice time or try to bolster false self-esteem through stubborn "victories." Each result simply becomes the next situation that stimulates more creativity and problem solving. Peace of mind need never be disturbed by any result. In fact, the more we maintain our inner peace no matter what situation we face, the more we are able to follow our version of the APPEARE process.

The perfect example of this is aikido, one of the Japanese self-defense arts. This discipline trains body, mind, emotions, and spirit to blend with the attacker's energy in ways that neutralize the attack. Judo has a similar philosophy. Because aikido does not teach attack or strike movements, the key to successful aikido, especially when one is being attacked by many people at once, is to maintain perfect awareness while moving in totally fluid (relaxed) movements.

If, when under attack, aikido practitioners allow either fear or vengeance in themselves, their total awareness is lost and they tighten up. Maintaining the constant inner peace while blending/asserting/acting in the real world is the art of aikido, and of creative action.

This integration of action and awareness can be seen as the classic masculine energy (logical, linear, action-oriented, outer-directed, mental, and so on) in its most productive form combined with the classic feminine energy (intuitive, artistic, awareness-oriented, inner-directed, grounding, emotional, and so on). It is the true goal of our physical, emotional, mental, and spiritual urges for optimum growth and health.

This same state is optimum for being productive in our work and with our co-workers. The harmony of action and awareness, the masculine and feminine principles, leads to our personal and organizational health and prosperity.

Evaluate Your Results

The president of Intel Corporation, Andrew Grove, has stated that we prevent work satisfaction for some people by divorcing them from the output of their work; then they usually play unproductive games. Others have had jobs where the work seemed endless—a constant flow of projects with no ending points to appreciate. And how many times have you heard the cry, "I never hear about the good work I do, but make one mistake and I never hear the end of it."

What happens to creativity in either situation? For most of us, the creative process needs a point of completion where we truly experience our results, even if they are not what we expected and/or wanted. Sometimes we may want the review and recognition of those results to be public, including praise and congratulations from others.

During any plane flight there is a constant interaction between the piloting and navigation, whether done automatically or in person. With all the variations of winds, pressure areas, and other factors, a plane must constantly make midcourse corrections. For even as much as 95 percent of the flight the plane is to some degree off course! Yet the sum of all the midcourse corrections results in a successful landing. Getting to your dreams, your goals, and your visions often demands the same process of self-correction.

A part of "results evaluated" is the formal performance appraisal process. Many people face these times with feelings of dread—feeling unappreciated or threatened by potential criticism or "unfair" gradings. Managers seem to dislike evaluation sessions as much or more than their subordinates. Yet the creative process demands answers to the questions, "Has the vision been realized? What are the results? What still needs to happen?"

Sometimes you may have failed to set up ways to acknowledge the end points to be found in a steady, endless stream of work. This can be even more true in "nebulous" areas such as creativity and innovation. Innovation objectives can be stated in two ways: (1) specific objectives for innovative ideas or projects, or (2) realistic "stretch" objectives that require innovative approaches to accomplish them. Success can be measured in terms of how well you foster an open, questioning environment for ideas; develop and document new ideas worth testing; and accomplish positive results using documented, innovative approaches. Your imaginings determine how you can set up constructive feedback for yourself.

IN CLOSING . . .

■ *I don't know how or where my ideas come from. It's almost an uncomfortable feeling, but I'm comfortable now with discomfort; . . . the creative feeling is, "I didn't really do this."*

I believe that everything that can be already is. All you are is a facility for creativity. What we create already exists but not in solid form. I conceive a product instantly as a whole thing, not in bits and pieces. The moment I've got it, I've got it; I know where every piece goes.

To stimulate others I feed the environment with tons of information. I pin up articles with comments like, "Critical development in plastics technology." Someone invariably comes up and asks why. One of my roles is to question why we are doing things—to satisfy our own ego needs? Client-dictated needs? Simply because it can be done?

John, director of design for a products design firm ■

Your creative process is a series of divergent and convergent stages, expanding and consolidating your thoughts, inspirations, and conclusions at every step. As the poet Alexander Pope reminds us, creating takes persistence and commitment:[2]

Fired at first sight with what the Muse imparts,
In fearless youth we tempt the heights of Arts,
While from the bounded level of our mind
Short views we take, nor see the lengths behind.
But more advanced behold with strange surprise
New distant scenes of endless science arise!

. . . .

. . . we tremble to survey
The growing labors of the lengthened way,
The increasing prospects tire our wandering eyes
Hills peep o'er hills, and Alps on Alps arise!

True ease in writing comes from art, not chance,
As those move easiest who have learned to dance.

True innovation also comes "from art, not chance." In chapter 1 I've mentioned that you can experience four creative *edges* where you work:

1. Promoting your organization's leadership in its chosen product(s) or service(s)

2. Challenging the frontiers of marketing, sales, technology, or other realms

3. Taking an active part in change

4. Exploring and developing your inner inspiration and creativity

Since then you have seen acronyms that describe three of the four key ingredients for fostering more innovation where you work. The four *faces* of the pyramid symbolize these key ingredients of your creative edge:

1. The CREATIVE climate for innovation

2. The SPIRITED people you work with

3. The APPEARE process of creativity

4. And *you*

When you act as a SPIRITED individual, using your version of the APPEARE process, you add to the aliveness, productivity, and success of your organization. You help yourself and others renew the organization, keeping it competitive and helping it serve ever-changing needs in the community.

You also help create the innovative organization described by Tom Peters, co-author of *In Search of Excellence*. In that classic book, Peters and Waterman described various attributes of the excellent, innovative organization, including the abilities to get new ideas into the marketplace quickly, to be close to their customers, and to operate by a few, shared key values. When asked to give the "bottom line" foundation for all eight, Peters said it was *"love and passion"*: ". . . loving what you're doing and doing it with passion." That passion means having the courage to hold and act on a positive vision of your future, investing your creative energies to bring it about.

Our workplaces are the perfect place to practice this creativity and courage. Directing our energies to fulfill our personal and organizational purposes and visions, we may initially be motivated by outside forces. Ultimately, however, our motivation must come from an inner sense of purpose, hardiness, peace, self-assurance, and *love*.

Why love? Because as social beings—as cells in the body of humanity—we find that serving ourselves and serving others are really the same. Deep inside we know that expressing our talents, serving ourselves, and serving others are identical creative processes. Likewise, harming others and harming ourselves are the same.

Love is often an unfamiliar emotion in work situations, but related feelings of *bonding*, or being close-knit, and comraderie are well known. At a time when many people have much more loyalty to their professions than to their organizations, bonding and closeness have become all important to maintain high productivity with low turnover. People will leave jobs to find a place where they can know a close-knit feeling.

In the past few years I have begun to see how each work situation I encounter can help me develop more wisdom, love, peace, and integrity in my life. I am moved—to feel, to be productive, to create—more easily now, with fewer reservations. I see that my desire to create and to express myself springs most deeply from a seedling, innate love of feeling alive. I've come to see the tragic creators—the Van Goghs and others who create in despair and passion—as struggling to reclaim that aliveness through creating, trying to pierce through their personal experience of a dark day-to-day existence.

From this I've concluded that the experience of creating *is* the experience of aliveness. Our aliveness ultimately includes our experience of connection, of love, of making a productive contribution, and of concern for the highest benefit of all. Creativity is more than an act or a skill and more than a style of working.

Creativity is a way of being and of profiting society (including organizational "societies"). Our work is one special place for us to make that contribution.

CHAPTER 5

Using Linear Techniques for Idea Generation

A Matter of Fluency

> *I almost overanalyze a problem from a logical structure—What are the pieces, what are the implications of this and that, what are the relationships? It's a very three-dimensional puzzle. The most creative example was a market research study I did for an amusement park on how many people might visit a new attraction. The problem in doing the project was that you can give any fraction of visitors you want, and "whoopee" you have any answer you want.*
>
> *We tried to figure out a way to structure it so that you couldn't give the "right" answer. I came up with a way involving the fraction of the people going to attractions, the average number of attractions that people visit, and people's ranking of attractions. I just sat down and played with the equations and it all fell in place. The result was within one percentage point of eventual, actual statistics.*
>
> *Bruce, litigation analysis consultant*

Sometimes policy and precedent reach their limits. Old solutions to problems may run their course, or the problems may simply be too different. New products, new ways to reach the public, new growth strategies, new operating systems become essential for revitalized growth.

Often the hardest part of generating new solutions is knowing where and how to begin looking for them. Whether you're seeking to create new material artifacts, new concepts, new spontaneous happenings, new events, new organizing principles, new modes of relationship, or new inner experiences, it's often useful to have structured ways of journeying to discovery. This chapter and the next can help you enhance your CREATIVE climate for innovation by focusing on techniques for using "Intuition and Logic."

A wealth of literature over the past fifteen years has described the functions of the two sides of the human brain. The left side has more to do with our linear, logical, mathematical, and verbal abilities. The right side has more to do with our intuitive, spatial, emotional, and musical abilities.

Both linear and intuitive thinking are necessary for optimum creativity. Most, and perhaps all, of the great scientific discoveries have occurred with the linear providing fertile ground for the intuitive to produce insights. For example, the German chemist Friedrich Kekule worked for years to discover the molecular structure of benzene. In experiment after experiment he gathered and analyzed information that might offer a clue. One night in 1865 he dozed in front of his fireplace:[1]

> *Again the atoms were juggling before my eyes . . . My mind's eye, sharpened by repeated sights of a similar kind, could now distinguish larger structures of different forms and in long chains . . . Everything was moving in a snake-like and twisting manner. Suddenly, what was this? One of the snakes got hold of its own tail and the whole structure was mockingly twisting in front of my eyes. As if struck by lightning, I awoke.*

He had envisioned the closed-ring structure of benzene. Following his insight, logic was again needed. It would have been a barren discovery had he been unable to formulate the insight and effectively communicate it to the world.

This is a typical pattern: logic preceding and following intuitive insight. Indeed, although intuition often embodies the glamor and mystery of the creative process, linear thinking is also essential. One way to view the link between logic and intuition is as follows:[2]

> *If someone said that marbles are like oil, you might think he had a few marbles in his head . . . (But) envision lubricating oil as millions of tiny marbles sandwiched between two surfaces. Suddenly it all makes sense.*

> *Most often our minds work logically . . . All creative ideas are logical when you view them in the abstract . . . Divergent ideas might not fit together very well, but take them to a higher level of abstraction and suddenly you've got a whole new concept.*

The essential difference between linear and intuitive thinking is that the former is sequential, whereas the latter is holistic. Idea-stimulation methods can also be classified in these two ways.

Linear approaches provide a structure within which we can seek and find alternative solutions. We arrive at solutions using a logical pattern or a sequence of steps. When we see our solution, we almost nonchalantly think, "Of course . . . there's the solution."

Intuitive approaches rely on a single image or symbol to provide a "whole" answer all at once. With the intuitive approaches we arrive at solutions by a leap. There is little or no experience of a path as the solution comes. When we see our solution, we often feel surprised, wondering, "Where did *that* thought come from?"

The ability to use both linear and intuitive approaches can (and often must) be drawn upon in our creative efforts. The overarching skill is to be *fluent* in both types of techniques. This chapter and the next introduce you to various linear and intuitive techniques I have found especially useful.

You can use both kinds of techniques either by yourself or in a group. They can be very powerful either way. In this chapter some examples are from individual use and some from group use to give you a feel for their flexibility. In chapter 9, on group problem solving, I give you other guidelines for using the techniques with groups.

Linear Techniques

Linear methods take advantage of different ways of organizing known information to help you approach problems from new angles. They help focus your attention on *where* to look for innovations, often the key to finding the optimum solution(s). For example, if you can determine from the beginning that your creative solutions will fall into any of x categories, you can focus on exploring just those categories. By proceeding from specific starting points, you can take small steps from one idea to another until suddenly you are far from the original starting position. In this uncharted territory you may find the innovative solution you desire.

There are many linear techniques available. These are ten that I have found very useful in business problems:

1. Matrix analysis

2. Morphological analysis

3. Nature of the business

4. "Reframing" questions

5. Force field analysis

6. Attribute listing

7. SCAMPER

8. Alternative scenarios

9. Forced or direct association

10. Design tree

Matrix Analysis

Imagine you want to develop some new product ideas. After having identified various market needs, available technologies, and product functions (what the product *does*) you might develop a two- or three-dimensional matrix to identify where to explore for new ideas. For example, you might use a market-technology matrix, as shown below.

		MARKETS			
		A	**B**	**C**	**D**
	1				
	2		*x*		
TECHNO-	**3**				
LOGIES	**4**				
	5				

Every intersection (*x*) represents a place to look for new innovations that apply a particular technology to a particular market. For example, to develop a packaging product for a plastics firm, your matrix might be very similar to this example:

		MARKETS			
		Transport	**Medical**	**Beverage**	**Industrial**
	Co-extrusion				
	Resin blend		*x*		
TECHNO-	**Laminates**				
LOGIES	**Adhesives**				
	Thermoform				

Perhaps you would use a market-functions matrix instead. For the purposes of idea stimulation, you would assume that you could develop or find any technology needed. Your market-functions matrix might resemble this:

		MARKETS			
		Transport	Medical	Beverage	Industrial
	Damage resistance				
	Moisture barrier		*x*		
FUNC-TIONS	**Reusable**				
	Lightweight				
	High temperature range				

Within each matrix you could explore new packaging ideas and even fill in the possible technologies to be applied, thus making a "working" three-dimensional matrix.

Morphological Analysis

This is a fancy title for a very simple and convenient way of generating solutions to problems that have many variables to consider.

For example, suppose you want to invent a new mode of transporting people and things. You must take into account:

- the driving force,

- the mode of movement,

- the material used in construction,

- the primary purpose.

Within each of these issues there are many alternatives to be considered. For example, under "driving force" you could list these possibilities: diesel, gas turbine, steam, pedal, electric motor, squirrels in a cage, horse, and wind. And under "mode of movement" you could list: rail, wheel, rollers, coasters, air cushion, water, and so on.

If you make a similar list underneath each of the issues, you get a table similar to the one on page 69.

You can develop new ideas by combining anything from the first column with anything from the second plus anything from the third and fourth. For example, how about a steam-driven vehicle that rides on coasters, is made of stone, and is used for carrying heavy freight? Or how about a squirrel-driven vehicle that floats on an air cushion, is made of plastic, and is used to transport spices?

The ideas that emerge from this method can range from the very

DRIVING FORCE	MODE OF MOVEMENT	MATERIAL	PURPOSE
diesel	rail	plastic	people
gas turbine	wheel	metals	animals
steam	rollers	wood	heavy freight
pedal	coasters	stone	baggage
electric	air cushion	cloth	foods and spices
squirrels in a cage	water	glass	plants
horse			
wind			

practical to the very outlandish. Although some of the ideas from this method will be eliminated after the first evaluation, they certainly open up the alternatives to consider—alternatives that might otherwise be overlooked. And even if the particular option isn't the correct solution, it may provide the stimulus for someone to devise the winning idea.

This method can work with any number of issues or variables. For example, imagine you wanted to find new opportunities in the food industry. You would deal with many food issues, including forms, kinds, properties, processes, and packages.

Some of the particulars of these issues might include:

FORMS	KINDS	PROPERTIES	PROCESSES	PACKAGES
preserves	meat	cost	ferment	bottle
drink	vegetable	convenience	freeze dry	can
chips	fish	nutrition	compact	pouch
flake	fruit	taste	blend	foil/paper
stew	dairy	texture	form	aerosol
roll	grains	odor	fried	box
soup	nuts	viscosity	bake	cup
topping	spices	medicinal	stir	sack
snack				

Pick one or two items from each list to make a complete idea. How about a soup made of fruit and spices with medicinal properties (antihistamines?), packaged as a freeze-dried product in small sacks?

A primary benefit of this method is that it conveniently structures the search for creative solutions of complex problems in a logical way that is easy to follow. Besides, it can be a lot of fun.

Nature of the Business

How the people in an organization define the nature of their business can have a tremendous impact on what they do and how they do it. When the Southland Corporation realized that their 7-11 stores were not really in the "grocery" business but rather the "convenience" business, it led them to an entirely different marketing strategy and inventory for their stores.

Businesses can define and organize themselves in many different ways, according to:

- their products or services ("a steel strapping company"),

- the markets they serve ("products for the transportation industry"),

- the functions they serve ("products for shipping stabilization"),

- their technologies ("products based on steel and polymer technologies").

For example, a bank could be defined as being in the financial business, the management-assistance business, (with financial resources to help implement management decisions), an information-processing business, or the achievement-and-experience-development business (with resources that enable a person to achieve his/her own entrepreneurial firm or to experience world travel).

If you were to work in a bank, its business definition might give you very different ideas for products and services. For instance, as a financial business you might offer loan packages and trust investment services. As a management-assistance business, you might offer software packages for a client's financial decision making or management consulting on corporate acquisitions. As an information-processing business, you might offer market-analysis services tailored to a specific company's target markets. As an achievement-and-experience-development firm, you might operate a travel agency in the banking offices or develop small business education packages (perhaps as computer-aided instruction).

Virtually all businesses are going through some redefinition of their business during this decade. For example, is a company that pays for health-related expenses in the *insurance* business or in the *health-care coverage* business? Depending on the answers, organizations in the health-care industries—from insurers to hospitals to pharmaceutical firms—are supporting or fighting different payment schemes, trends for self-insuring businesses, expansion of health-maintenance organizations

(HMOs), proposals for government involvement, and so on. The stakes are high and getting higher as health costs escalate.

Building on this technique and using the morphological analysis method just described, you could explore the nature of your business by combining individual opportunity ideas in multiple ways. For example, you might want new plastics packaging products, but what really is your business? Your variables would be markets, functions, technologies, products, services, and process equipment. You could first make a *key-word* index as seen below.

MARKETS	FUNCTIONS	TECHNOLOGIES
beverage	damage resistance	adhesives
medical	moisture barrier	laminates
industrial	reusable	coextrusion
transportation	lightweight	thermoform
toys	high temperature	resin blend

PRODUCTS	SERVICES	PROCESS EQUIPMENT
cans	leasing	case opener
cups	manufacturing supply	lidding machine
trays	repair	wrap machine
pouches	training whole systems	depalletizer
cartons		conveyor

Every new idea could then be expressed in terms of these key words. One idea, for example, might be a damage-resistant, thermoformed can for the toy industry, with case operation equipment and manufacturing supply services also provided.

With all ideas expressed in terms of key words, a computer can sort all the ideas for a given item, for example a business built around the beverage market or damage-resistant functions or adhesive technologies. This type of ordering makes it easy to cluster the opportunities to show overlaps in market penetration, to demonstrate the viability of technology investments, and so on. From these insights you can examine the possible primary definitions of the type of business that would take advantage of your best opportunities.

Reframing Questions

Reframing is a technique in which you ask questions in order to find new viewpoints in understanding a problem. With the new viewpoints you can develop different definitions of the problem and generate ideas for each different definition.

It has been said that Einstein's genius lay partly in his inability to understand the obvious. Similarly, in tackling your own problems, try going beyond your normal frame of reference. Break your typical patterns of thought. This will help you find the best possible outcome rather than just solve the obvious or immediate problem.

Some of the questions you can use include:

- What is the broadest frame of reference for this problem?

- What are the "givens" here—the "obvious" realities of our situation—and how can each be challenged?

- What are typical complaints (pet peeves of customers, etc.)?

- What is the ideal state of affairs we are looking for?

In some ways, these questions can put you further outside the problem to get a new look at what the problem really is. In other ways reframing questions will put you further into the middle of the problem. T. A. Rich, a famous inventor at GE, "put himself in the middle of a problem; trying to think like an electron whose course is being plotted or imagine himself as a light beam whose refraction is being measured."

Force Field Analysis

This method was first developed by Kurt Lewin, a social psychologist. Its name comes from its ability to identify forces contributing to or hindering a solution to a problem, and it can stimulate your creative thinking in three ways: (1) to define what you are working towards (your vision), (2) to identify strengths you can maximize, and (3) to identify weaknesses you can minimize. The method is quite simple.

Using a format like the one that follows, pick a situation you would like to see changed—for example, your product line, work conditions, or relationship with your boss.

1. Write a brief statement of the problem you wish to solve (write objectively, as if you were a newspaper reporter).

2. Now describe what the situation would be like if everything fell apart— absolute catastrophe.

3. Now describe what the situation would be like if it were ideal.

4. Presume the center line represents your current situation. "Catastrophe" and "ideal" are playing tug-of-war. Fill in what forces are tugging *right now* at your situation to help make it more ideal and what forces are trying *right now* to make it more catastrophic.

Suppose you want to explore how to foster a better climate for innovation and creativity in your organization. Using observations from the example in chapter 2, your force field would include the elements in this analysis:

FORCE FIELD ANALYSIS

(1) Problem: How to foster a better climate for innovation *and* creativity.

(2) Catastrophe: No innovation in products or marketing

(4) Forces

(3) Optimum: High innovation with responsible risk

−	+
A person with a new product idea not knowing where to take the idea if his/her supervisor doesn't like it	A new product succeeding in the market because of up-to-date information on consumer values and life-styles
A brainstorming meeting producing uninspired or impractical ideas or ideas that don't get acted on	An individual finding time to incubate a problem, literally "dreaming up" a solution
Performance appraisals and rewards that encourage short-term profits and status quo operations at the expense of long-term investment in needed innovations	Senior management clearly stating a vision of the organization's future and the requirements for market-oriented, innovative new products and services

The primary function of the force field in idea generation is to present three different stimuli for thinking of new options or solutions. Because the field represents a kind of tug-of-war, there are three ways to move the center line in the direction of the more desirable future:

1. Strengthen an already present positive force

2. Weaken an already present negative force

3. Add a new positive force

Therefore, the force field presents you with focuses for exploring possible solutions. You can then employ matrix analysis or other idea-generation techniques. The primary benefit of the force field is that it identifies strong points in a situation as well as problem areas. These strong points can be the foundation of the most effective solutions, which might have been overlooked otherwise.

Attribute Listing

Whether you have a procedure, product, or process you wish to improve, one method of getting ideas is to write down all the attributes or components and see how you can improve upon any one or all of them.

For example, a bicycle has these attributes among others:

- frame
- pedals
- drive sprocket
- chain
- rear sprocket and chain guide
- brakes
- tires
- handlebars

Each attribute has seen dramatic improvements and innovations in the last thirty years, including:

- much lighter weight frames based on new, lightweight materials;
- pedal grips and straps to secure feet better;
- double-drive sprocket for ten gear capabilities;
- stronger chains with special clamps for easier changing;
- improved derailleur gears for rear sprocket;
- hand brakes that grip tires to replace rear axle brakes;
- racing handlebars for more effective racing position.

Attribute listing is similar to force field analysis. The force field provides specific negative and positive aspects of the problem, whereas attribute listing provides neutral aspects. Both identify categories in which improvements might be found.

Scamper

Alex Osborn, a pioneer in creativity facilitation, developed a list of "idea-spurring questions" that were later arranged by Bob Eberle as the mnemonic SCAMPER:[3]

S = Substitute? (other ingredients, material, power, place . . .)

C = Combine? (blend, alloy; combine purposes, appeals . . .)

A = Adapt?

M = Modify? (color, shape, motion . . .)
    Magnify? (stronger, larger, thicker, extra value . . .)

P = Put to other uses?

E = Eliminate?

R = Reverse? (roles; try opposites, upside down . . .)
    Rearrange? (pattern, sequence, pace, components . . .)

Apply these questions to your situation and see what ideas emerge. If you wanted to develop a new office procedure or work flow, you could first identify all the steps currently being taken. Looking at your list, use SCAMPER as a mind jogger to imagine many different ways to do the same work.

Many new products have been developed using opposites, or the reverse. King-sized cigarettes came from "short-long"; sidewall tires from "black-white"; powered car windows from "manual-powered"; and so on. Substitution has also been important. Milk cartons substituted paper for glass. Fiber-reinforced composite materials have been used in tennis rackets and airplanes to substitute for materials with inferior strength-to-weight ratios.

Alternative Scenarios

There are two primary ways to explore the range of possibilities for the future: hypothetical situations and alternative scenarios. Both are excellent in generating new approaches to business by breaking habitual ways of perceiving the environment.

With hypothetical situations you make up something and solve it: "If a particular set of conditions prevailed in my industry ten years from now, what would I do—then and now? To what conditions is my organization most vulnerable? What can we do in response to these vulnerabilities?"

Alternate scenarios are a more comprehensive way of asking these same questions. Scenarios are qualitatively different descriptions of plausible futures. They give you a deeper understanding of potential environments in which you might have to operate and what you may need to do in the present.

Since the early 1970s companies such as Royal Dutch Shell have used alternate scenarios to make decisions that would position them for sudden discontinuities in the normal trends. For example, in 1971 one of their four scenarios contained a description of a sudden oil shortage and a sudden high rise in oil prices.[4] Although they hadn't actually pre-dicted the 1973 crisis, they had broken out of status quo notions of how to position their business activities by strategizing across many possible futures. Partly because of this, they grew from number seven to number two in their industry during the 1970s.

When managers do long-range planning based on a single forecast of trends—a single notion of what their market and business will be like in two, five, or ten years—they are actually taking the risk of "betting the company" on that single forecast. Even with "high, low, and prob-able" projections, the first question usually asked about a forecast is, "Given its assumptions, how far off can it be?" Besides being risky, this practice can also prompt people to avoid action during uncertain times, paralyzing management decision making. Scenarios help you to identify what environmental factors to monitor over time, so that when the en-vironment shifts, you can recognize where it is shifting *to*.

Thinking through several scenarios is a less risky, more conserva-tive approach to planning than relying on single forecasts and trend analyses. It can thus free up management to take more innovative ac-tions.*

Rather than being general "future histories," scenarios are devel-oped specifically for a particular problem. To begin developing scenar-ios, first state the specific decision that needs to be made. Then identify the major environmental forces that impact the decision. For example, suppose you need to decide how to invest R & D funds in order to be positioned for opportunities that might emerge by the year 2000. The major environmental forces might include social values, economic growth worldwide, and international trade access (tariffs, etc.).

Now, actually build four scenarios based on the principal forces. To do this use information available to you to identify four plausible and qualitatively different possibilities for each force. Assemble the alterna-tives for each force into internally consistent "stories," with both a nar-rative and a table of forces and scenarios: Build your scenarios around these forces. Each scenario will have a story line.

A midwestern bank used scenarios to stimulate new ideas for main-taining a strong consumer-lending business in upcoming deregulation. Scenario story lines emerged for "Business Today to 1990," "Heated," "Belt Tightening," and "Isolation." The chart found on pages 77 and 78 details each of the bank's scenarios (some driving forces have been omit-ted):[5]

---

* Curiously, many Japanese and European firms seem very open to using alternate scenarios. American companies often think that planning more than three to five years ahead is somewhat impractical and that forecasts are suitable for such time periods.

DRIVING FORCES

SCENARIOS

DRIVING FORCES	BUSINESS TODAY–1990	HEATED	BELT TIGHTENING	ISOLATION
ECONOMIC:				
Interest rates	Moderate–high, 9–12%	High, 11–15%	Moderate–low, 8–10%	Low, 5–9%
Economic health	Relatively slow growth	Rapid growth, inflation, 7–8%	No recession/less disposable income	Recession
U.S. foreign trade	Deficit	Balanced	Deficit	Deficit
Third World debt	Confidence in Third World	Confidence in Third World	Third World okay, with restrictions	Concern high
U.S. budget	Deficit/no panic	Deficit	Balanced	Deficit
Protectionist legislation	No legislation/none of significance	Mutual trade	Low–moderate legislation	Moderate–high legislation
TECHNOLOGICAL:				
Imaging/internal process	Read only	Interactive/read	Interactive	Read only
Cable/satellite	80%	90% (20% 2 way)	70%	50%
Home computers	60% affluent/30% general	70%/35%	50%/25%	40%/20%
"Pacing" technology	No major breakthroughs	Breakthroughs	Breakthroughs	No major breakthroughs

DRIVING FORCES

	BUSINESS TODAY–1990	SCENARIOS		
		HEATED	BELT TIGHTENING	ISOLATION
COMPETITOR:				
Consolidation of industry	High	Moderate	High	Moderate (federal controls); moderate number of failures
Intensity	Very intense	Intense	Intense	Less intense
Options (relative)	Many	Many	Some	Few
Technical usage	Not significant	Major applications widespread	Few applications in a concentrated number of companies	No major advantages

In Royal Dutch Shell's 1975 scenarios, for example, two story lines described possible recoveries from the 1973 oil crisis:

1. "Boom and Bust" described a vigorous recovery, more characteristic of the fifties than the sixties, that would end in its own destruction.

2. "Constrained Growth" described a muddling through that would be more halting and slower than previous business recoveries.

With the scenarios in hand, identify business opportunities within each scenario. Then examine the links and synergies of opportunities across the range of scenarios. This would help you to formulate a more realistic strategy for investment.

Using these scenarios helps identify what environmental factors to monitor over time. When the environment shifts, you can recognize where it is shifting *to*.

Forced or Direct Association

When you put two concepts together that seemingly have nothing in common, you might be very surprised to see what ideas emerge. In a workshop to identify new ideas for constructing walls, participants were asked to make a connection between wall construction and spiders. One idea generated was to manufacture an adhesive webbing that would make it easier and faster to put up inside wallboard. Another idea was a new design for a caulking gun.

Or suppose you are wondering, "How can I improve my relationship with my boss?" You can ask, "How is this relationship like a pencil?" If you imagine a yellow pencil with an eraser, you might think . . .

- eraser . . . We both keep bringing up past mistakes.

- lead . . . I procrastinate. (I need to get the lead out and confront this.)

- yellow . . . I feel timid.

- gold ring . . . He's doubting my commitment to work.

- wood shaft . . . I'm getting shafted by taking on too much.

From this thought process, you can identify many ways to respond to the situation.

Design Tree

The design tree is a type of "mind map" that is useful when you have a central topic—a product, market, technology, or process, for example—that you want to build on. For example, suppose you want to build a new product line around your great water softener product. Your "tree" might begin something like this:

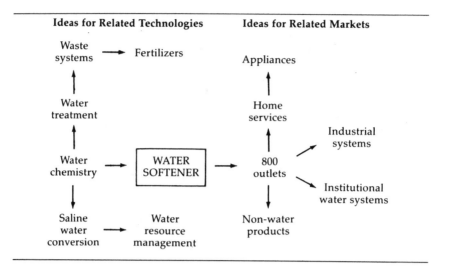

Ideas for Related Technologies	Ideas for Related Markets

Notice that the design tree is a mapping of ideas that occur to you based on related technologies and markets. With other problems the branches may be based on personnel, production, or other considerations. As you can see, the drawing can look confusing to an outsider, and the relationships between ideas may not be obvious. But these types of maps parallel the way our minds work sometimes. Doing this exercise with a large piece of paper on the wall can get you away from obvious ideas and onto potential breakthroughs.

IN CLOSING . . .

> In my discipline, instructional design, one moves towards a goal, an acquired skill. I think of instruction as a bridge. I really feel myself using whatever I've got in my tool kit to help me build that bridge. Creativity within the instructional process really comes down to manipulating the different ways one can engender strategies, and even within a strategy, the choice of approaches and how one phrases one's words.
>
> John, educational technologist

> When I have a research problem, I usually don't come up with just four approaches. I may come up with fifteen, and it's impossible to make a reasonable evaluation . . . Usually you can throw away half of them, but (with) the other seven, say, the problem of defining the approach becomes complex enough that

*you can't settle on any of those. Then you have to sit on it and
. . . it congeals, shall we say.*

—Rick, lab director in organo-metallic chemistry ◀

There are many other linear techniques to stimulate your creative problem solving and to give you a clearer path for your follow-up actions: Kepner-Tregoe, value engineering, Delphi, and mathematical analyses of data, just to name a few. As mentioned earlier, all of these can be employed by you individually or as part of a group.

Either way, the linear techniques are very powerful in showing you *where* to look for innovations. They are complemented by the power of the intuitive techniques to tap your inner source of creative insight. That is the subject of the next chapter.

CHAPTER 6

Using Intuitive Techniques for Idea Generation

Of Higher Rationality

🖢 *I'm valuing more and more the whole area of visualization, or imagery techniques. When I'm talking with a person about his/ her skills, his/her values, and his/her interests, sometimes I will get an image, a picture of something. Often, I feed this back to the person and ask, "What do you think? Does this make sense?" It will clarify something. It gives him/her a fairly clear picture of something they weren't able to put words on.*

I recently was working with an incredibly brilliant woman who is getting to be nationally famous in her field. She could out-logic anybody in the world. In career counseling we went through skill analysis and values, all this analytical stuff. She came back and said, "Well, so, I'm still not clear about what it all means."

I said, "Well, let's take all that analytical stuff and put it away. Just close your eyes and reflect a bit." It worked into a visualization about going and talking to an inner advisor. I led up to it by having her talk to different animals, plants, inanimate objects, getting messages from what they "said" to her. She saw herself as a weeping willow in one case, and that really had significance for her. She was absolutely blown away. It gave her clues that she wouldn't have gotten from logical analysis.

Diane, career counselor at a research facility 🖣

Many of us have been raised to consider the inner world of our experience to be too unreal and untrustworthy, too full of imagination and emotion. Many executives who are excellent in using their intuition to help make decisions refuse to admit it. While using facts and logic to provide the foundation for their decisions, they hide their actual decision-making processes, dressing up their intuitive decisions in "data clothes."

Your intuitive insight is often prepared by the persistent, linear work. The logical *without* intuition and emotion is actually irrational, because it is not based on our full capacity for problem solving. Because logic alone is not whole, it cannot produce whole, reality-based solutions or promote the integral, long-term health of any person or organization.

The realm of intuition is *not* unreal or untrustworthy. Our intuition remembers data that our conscious mind has stored away. It is our inner, intuitive world that is constantly giving us the guidance and answers to our questions about living and problem solving, *especially* when our logical, linear thinking reaches its limit.

Linear techniques structure information and point out where to look for new ideas. Intuitive techniques take advantage of our right-brain capability to perceive whole solutions in sudden leaps of logic. Our intuition is more fluent in images, sounds, and symbols than in words—as in our day and night dreams. Intuitive techniques take advantage of the superior insight often available in these images, sounds, and symbols.

An important assumption—or truth—to hold when using intuitive techniques is that at some level *you already know the answer.* You already have that which you are seeking. It is as if you had misplaced your favorite jewelry somewhere in your top drawer; if you keep looking, you will eventually find it. This is a different perspective from, "Is there an answer?" or "Will I ever solve this?" The assumption that you *already know* can open the main door to the solution.

To tune in properly to our intuition, it is important to realize that our intuitive self is *not* the same as our subconscious. Our subconscious includes memories and related thoughts and emotions, which may actively affect our day-to-day life. Our intuitive self comes through when such subconscious rumblings have been quieted. We can make much greater use of our intuitive wisdom if we are willing to nurture inner calmness; set aside our inhibiting habits, beliefs, and emotions, and really listen to and dialogue with our intuition.

There are many methods for developing this inner calmness and tuning in to your intuition. For example, you can use the strong link between breath and mind. When your body is relaxed and breathing slowly, it is difficult to be madly thinking or emoting. *Following* the breath (not controlling the breath) is a major focus for stilling your mind to allow deep, intuitive insights to come forward. This chapter contains an exercise to experience this firsthand. Many other methods can be found in stress management literature.

Intuitive Techniques

There are six intuitive techniques I have found most useful in solving work-related problems:

1. Imagery

2. Brainstorming

3. Analogy

4. Dreams

5. Drawing

6. Meditating (mind-clearing)

These are all effective means of using your intuition and evoking insight.

Imagery

Imagery means using symbols, scenes, or images as windows to intuitive creative thought. A circle, a gurgling stream in a fragrant meadow, or a snake eating its tail may somehow hold the key to the solution you're looking for. Imagery is a vehicle for communicating qualities, bringing paradoxes together, and expressing the meaning we ascribe to information and experiences.

The right brain communicates in images rather than words. You can dialogue directly with your intuition through images and then abstract the qualities of the images into words. This dialogue can be in the form of asking questions and looking inside for whatever image comes up.

For example, I once worked with executives from a department store chain and a savings and loan association. They wanted to identify the corporate values for operating a new joint venture. After discussing demographics and other logical data, I had them close their eyes, relax, and ask their intuition for images that symbolized the key "corporate culture" values for making their joint venture a success. One person imagined a big family dinner; another saw herself talking with a close friend; another saw himself driving a sports car on the California coast. We then discussed the qualities found in these images, such as respect and service. They formulated their own meanings for these values, which became the key principles to communicate with the public and to practice with their employees.

This shows the combined power of physical relaxation and intuitive imagery. It also shows how the imagery can be nourished by logical analysis but can give deeper answers than logical analysis alone can provide. Imagery is not independent.

There are a number of guidelines for using imagery:

a. Relax. Soothe both body and mind using breathing or another technique.

b. Clearly ask your intuition for an appropriate image. (e.g., "Give me a symbol, scene, or image that represents a new advertising theme.").

c. Accept whatever images emerge. Surrender. (Don't judge or sort through, take what's there.)

d. Make the image vivid. Encourage emotions and many sense modes (feelings, sights, sounds, smells, tastes).

e. If the image is not easy to understand, ask your intuition for another one.

f. Honor all intuitive messages. Enjoy them.

g. Look for the *qualities* in the image (rather than getting caught up in the literal meaning).

The more fully you create an experience in your mind—with clear, multisensory images—then the more powerfully can your intuition communicate to you. Sometimes you may wish to augment the imagery process with musical selections.

When you get an image, talk to it and see what it says back. If you receive an image, for example, of a red sports car speeding past a luxurious old ranch home on a road to Las Vegas, you can ask each of these symbols to tell you what they are doing in your image.

At first, images in your mind may be hard to understand, or you may not like them. If you get only partial information, you can dialogue to fill in the gaps. If an image is confusing, ask for another image that is easier for you to understand. Your intuition will gladly oblige, especially if you say, "Thank you," for all images it offers you. Your intuition is a friend of yours. It's a part of you.

If an image seems frightening or disturbing, keep asking for the next "deeper" image. Finally one will emerge that you can work with fruitfully.

As we learn to trust our intuition more and more, we are able to find and accept the truth inherent in the images as soon as they come up. Thus, our vision and insight become more powerful and creative, leading to more effective innovation and personal expression.

A team of specialists from a Japanese software company was working with consultants to identify new business opportunities in software. As part of a three-day workshop, I led them on an imaginary ocean dive, seeing fish, picking up artifacts, and finding a treasure chest. When we discussed what they had imagined, the qualities inherent in their images spawned many new ideas. Wet suits gave us ways to "warm up" the user interface with a cold technology. An octopus gave new ideas for developing the "architecture" for elaborate computer systems in artificial intelligence.

Try a similar guided fantasy with a situation of your own. It will help you appreciate and understand your abilities to work well with imagery. If you're like me, you might prefer to read on rather than do an exercise; I get comfortable about absorbing information and often don't like to stop to participate actively in the learning. Yet many of the ideas contained in this book cannot be truly learned until they are experienced inside. You can't get filled up by reading about a gourmet meal; you have to eat it yourself. You can't experience skydiving or sex or relaxation or creativity just by *reading* about each in a book.

After reading this exercise through once, close your eyes and actually take a few minutes to experience it. If you prefer, have someone read it to you, or dictate the exercise into a tape recorder and play it back so you can more easily follow all the instructions.

First, take a few minutes to review the seven arenas for creativity: idea, material, spontaneous, event, organization, relationship, and inner experience.

Now think of a situation at work that needs improvement or special creative insight. Write this down in a sentence or two on a piece of paper as if you were a newspaper reporter giving a completely objective overview of the situation. Now close your eyes and lead yourself into a state of deep relaxation.

Imagine that you are on your favorite warm, sunny beach, perhaps in the Bahamas or on the Hawaiian Islands. Take in the whole scene with all of your senses. Feel the warm breeze blowing across your face and the smooth grains of sand massaging your feet. Hear the waves rolling gently onto the shore, and the leaves from the palm trees and grasses brushing softly against each other with the breeze. Smell the salt water and the sweet smoke of a campfire burning nearby. Look around and see the curve of the shore and the people with you on the beach. Several ships may be passing by on the horizon.

You put on your scuba-diving gear. You feel confident. Your scuba-diving lessons prepared you perfectly for a dive in this warm, clear water. You carry the single tank easily on your back as you walk to the water's edge, put on your flippers and mask, and insert the regulator into your mouth. The air draws easily and you breathe naturally.

Excitedly you wade into the waves until they reach your chest, then you dive under and begin to swim about ten feet below the surface. As you look around hundreds of beautiful tropical fish— glowing oranges, blues, and pinks—swim around your arms and legs and dart in front of you. You continue to breathe easily, swim further out from the shore, and dive a little deeper. You are still in very safe water. There are no dangers anywhere.

Then as you gaze down to the ocean floor, about fifteen feet below you, something captures your attention. You swim toward it. If it is alive, you trust it and it trusts you. No danger exists for you . . . or it.

Ask it to speak to you, to say how it represents a solution to your situation. Have a dialogue with it. Ask it to clarify anything you want to know. Test its insights against your own feelings and thoughts. Acknowledge its contribution to your thinking. Then thank it and swim on, knowing that you will remember everything it told you.

As you swim further, you notice below you a small, sealed chest nestled in the sand and surrounded by dark green seaweed. You swim down to it and open it. Inside you find a piece of paper folded in half.

As you unfold the paper, you see that on it is written the message in response to the work situation that needs improvement or special creative insight. You read the message. As you swim on you begin to contemplate how the message might be a solution to your situation. You test it against logical considerations.

You return to the beach, take off your scuba gear, and lie on the warm sand. The warm sun and warm breeze caress your skin.

You reflect on the solutions you have found, noting to what degree they solve your situation.

This is just one type of exercise that shows how you can involve all your senses in tuning in to your creative intuition. After all, no one wrote anything on that piece of paper but you, even though you may have been surprised by the message you received. This exercise also shows how intuitive images can be enhanced with linear, verbal messages, thus giving you fluency in both.

Brainstorming

This is perhaps the most well known—and sadly, the most abused—of all idea-generation methods. It is the most abused mainly in that many people try to use it either without understanding the essential ground rules or when other methods would fare so much better. Although it is usually considered a group technique, and the guidelines are phrased as such, it is also an important private method of idea generation.

In brainstorming state your problem and then give your top-of-the-head ideas in any order. Some ground rules for effective brainstorming include:

- Pick a problem each person can help solve.

- Define the problem in neutral terms rather than as a preselected solution (Use something like "How can I get this job done?" instead of "How do I get John to do this job?").

- Record the ideas on a flip-chart or large pieces of paper where everyone can see them.

- Suspend evaluation or judgment until all ideas have been given.

- Stretch for ideas.

- When you think you've gotten all ideas, go for another round, being even more outrageous in possible solutions.

- Aim for quantity to help find quality.

- Accept all ideas, even wild ones.

- Encourage embellishments and building on ideas.

In defining your problem it is important not to make it so narrow that it limits your imagination. This usually happens when you define your problem in terms of a specific solution. For example,

Instead of . . .	Say . . .
"How can we get that department to write the new procedures?"	"How can the new procedures best be communicated?"
"How can we build a better mousetrap?"	"How can we respond to our mice overpopulation?"

The "no evaluation or judgment" guideline frees up the more inhibited and shy to contribute. More importantly, it frees you from your internal "idea killers," those negative judgments you might immediately make about a new idea, judgments you might have a hard time turning off when other people are giving their best ideas or even when you are thinking of your own.

Some of the more frequently used idea killers are:

- ". . . but our business is different."

- "We've done OK without it."

- "Let's shelve it for the time being."

- "I considered that myself once, but . . ."

- "That's not my/our responsibility."

Q What idea killer do you hear where you work (make a list)?
     What idea killer do you most often use yourself?

Assume you can suspend your negative judgments for the time being
(you can take them back afterwards). You can spur your creativity by
changing idea killers into positive affirmations—statements that accent
a positive outcome. For example, instead of "That's not our responsibility," can you allow yourself to think, "I/we take responsibility to consider all possible solutions"?

There is a primary difficulty in using brainstorming: it gives no
guidance on where to look for ideas beyond the more obvious ones
found in the presentation of the problem. By contrast, morphological
analysis and force field analysis give more avenues to search for innovative solutions. However, you can combine brainstorming with other
methods—for example, with each question in SCAMPER or each element of an attribute listing. Brainstorming can thus play a powerful supportive role in idea generation and problem solving.

Analogy

An analogy is a similarity between two conditions or things otherwise
dissimilar. The use of analogies assumes that if two things are alike in
some respects, they must be alike in other respects as well. Analogies
serve to help "make the familiar strange and the strange familiar." From
either of these perspectives you can often find new approaches and new
insights into the nature and possible resolutions of the problems and
questions you are working on.

To use *personal analogy*, identify yourself with an object or process
in order to get a new perspective on the problem. For example, if you
were trying to invent a new typewriter, you might ask yourself, "If I
were the carriage return, how would I feel? What would I want?" Or if
you were trying to invent a new electronic printer, you might imagine,
"If I were a piece of paper, how would I like to have characters printed
on me?"

To use *direct analogy*, make a comparison between parallel facts in
two different fields to get new light on the problem. For example, if you
had been Alexander Graham Bell trying to invent the telephone, you
might have compared the eardrum and bones of the human ear and how
they work with the way some other membranous material might behave
with "bones" of metal. As another example, the inspiration for Velcro
came from observing how burdock burrs cling to clothing.

To use *symbolic analogy*, use an image that might be technically incorrect but that colorfully describes the implications of a key concept of
the problem. For example, to cover buildings where the outside surfaces

take a beating from the elements, your analogy might be "a paint that grows thicker with age."

Similarly, you can take the *purpose* of a solution and imagine ways to fill it: "———(Something) that ———(performs a function) like a(n) ———(analogy)." For example, you might want to invent "*new clothing fibers* that *change color* like a *chameleon*."

To use *fantasy analogy*, imagine anything without regard to real-world plausibility and see what real-world ideas can spring from the exercise. For example, if you were trying to invent new arthroscopic surgery devices (which are inserted into the body on thin wires to perform microsurgery without major cutting of body tissue), you might imagine being able to shrink people to go on a "fantastic voyage" through the body to make repairs.

Analogy and imagery go hand in hand. Analogy tends to be more directive in generating new ideas, because you start with a comparison point (e.g., the starting point for Velcro was the burdock burr). Imagery, on the other hand, is usually more open-ended, having no beginning point for comparison. Both are extremely useful on their own and in conjunction with the other intuitive methods and the structures of the linear techniques.

Dreams

Solutions can come to us when we don't expect them, perhaps after some "gestation" period in our minds. This method takes advantage of our wonderful ability for both daydreams and night dreams. Ray Bradbury, the science fiction writer, once said that he often discovered preciously good material in the half-awakened, half-slumbery time before real sleep. Often he forced himself completely awake to make notes on these ideas. You can coach yourself to receive dream images by asking your intuition, "Give me a dream about ——— (the problem you're working on). Awaken me as soon as the dream is over." (With practice, you will be able to dream and wake like this.) As soon as you're awake, *don't* open your eyes but *do* review your dream. Then open your eyes and get the pad and pencil you have left by your bedside and quickly write down the main elements of the dream.

The key is to be willing to write down the key images or words that come to mind as soon as you realize they are there. The insights can be hard to recapture, so take advantage of their first visit. Later (in the morning, for example), you can fill in gaps and other details you can remember, settings, feelings, sensory images, and qualities you can distill from it all. As in other interpretations of imagery, look for the qualities represented by the images, treating them as metaphors rather than taking them too literally.

This process also works well after relaxation, meditation, and just plain daydreaming. (Keep a pad with you in the car for insights while driving and in the bathroom for insights while bathing or shaving!) Daydreams can be invoked simply by creating a fantasy about some ideal world or situation you would like to be in. You can also develop a "wish list" of items in your daydream that would combine to solve the problem you're working on. You will then evoke either images or words that you can write down—perhaps even write them in your daydream—to help bring the insights into your waking consciousness. You may even find that you can play "movie director" for your daydream to guide the course of events partially.

Drawing

Because your intuitive consciousness communicates more easily in impressions and symbols than in words, drawing is a marvelous way to bring out creativity. Drawings can both *evoke* and *record* creative insight, and different techniques have been developed emphasizing one of these two purposes over the other.

Evoke   To elicit deeper insight, drawing relies on producing intuitive symbols, sometimes archetypal. One exercise to help diagnose a problem situation is to become deeply relaxed and ask your intuition for symbols, scenes, or images that represent your situation. (This can also be done as a guided fantasy, similar to the one given earlier). After getting the symbols, use as many colors as possible and draw the symbols on a large piece of paper as your *hand* wants to draw them. (You may wish to use your "opposite" hand to give you less conscious control over the drawing.) Then fill in the first word that comes to mind for each of the symbols. Write a paragraph combining all the words, and expand on those thoughts in a free flow of thoughts and feelings.

Another example of using drawing to evoke insight is to establish a *theme* for a meeting (for problem solving, strategy development, etc.). Over the course of a few days before the meeting, one person generates images that represent that theme. The meeting then begins with the image placed next to the agenda and objectives to elicit comments about the purpose of the meeting. This almost always elicits very inspiring principles and values. In turn, these values can energize and empower the group to embrace the theme and engage in the meeting wholeheartedly.

For example, a very successful social service agency held a meeting with the theme, "After Success, What? 1990 and Beyond."[1] The image that the facilitator developed before the meeting can be found on page 92.

As the participants discussed the image, the star came to mean their success, and the heart inside it was the source of their success. The road symbolized going on another journey toward mountains that represented new challenges. They emphasized that they didn't want to be-

AFTER SUCCESS WHAT? 1990 + BEYOND

come complacent; rather, they wanted to take on new risks. The sun nearby was their very dynamic past, and the more distant sun was their dynamic future.

As another example of how drawings can help conceptualize the nature of the problem you are working on, the drawing found below was developed in the course of a strategy conference.[2]

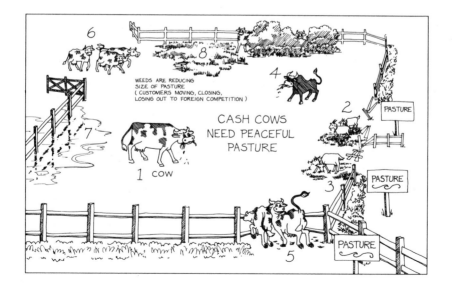

6

8

WEEDS ARE REDUCING
SIZE OF PASTURE
(CUSTOMERS MOVING, CLOSING,
LOSING OUT TO FOREIGN COMPETITION)

4

2

PASTURE

CASH COWS
NEED PEACEFUL
PASTURE

7

PASTURE

1 COW

3

PASTURE

5

PASTURE

The central cow (1) symbolized the company itself, and the other animals were competitors:

a. Competitors with more energy-efficient products (2)

b. Competitors with different high-quality features (3)

c. A competitor who produced very large-scale products (not direct competition) but who also was licensing its technology for smaller applications (4)

d. Future competitors who could enter the market using emerging computer capabilities (5)

e. Competitors with narrowly focused but high volume applications (single customers) (6)

The water (7) was the invasion of new technologies, while the weeds (8) symbolized the natural turnover of customers.

After this drawing was shown, the president of the company studied it for the remainder of the day. The next day he came into the meeting room excitedly saying, "I know what we're going to do!" He described a strategy of acquiring one cow to avoid competition, changing pricing strategies to fend off other cows, and so on. The image gave him a total sense of how to rejuvenate his business.

Drawings are also essential for conceiving many engineering and scientific innovations—witness the extraordinary power of computer-aided design and engineering programs to make drawing processes more elegant and time efficient.

Drawing can be used in less elaborate ways, too, to deliver insight. After a period of deep relaxation or contemplation, if you have had your intuitive self working on a problem for you, draw an image of the current state of your problem/answer. Ask yourself, "What is the current state *today?*" knowing that you don't need a final answer right away. Suspend judgment—especially those voices from age eight that said you weren't good at art in school—and use colors or pens or even magazine cutouts. Try letting the images flow without conscious direction, as if the items on paper were telling you how they wanted to be. Then put words to the images—whatever words come through your hand. Again, use the results as a "status check" to see what is in your subconscious, impressions that you can always modify if you want to.

Another method is first to write down a problem on a piece of paper and then to put it away. Use an imaginary language of nonsense sounds to write a four-line rhyming poem. For example, your problem might be, "Develop a new concept for single-family dwellings for young achievement-minded people." The poem might be:

Con dabble on onna nu
Wenp falli ton immu ma
Zel chima fon digi gu
Fen walli non willi la.

Now write a translation of the poem in English; the sentences do not have to follow each other logically. The "translation" might be:

Don't dabble with minor news.
Immense ideas will make competitors weep and pale.
A santa chimney gift: fun by single
Fenders and body walls that can be modified easily.

Then look back to the problem statement and see what suggestions you get from your translation to resolve it.[3]

Record There are two general methods of recording ideas as an individual or group generates them. The first is more linear, in traditional outline fashion:

I. _____

    A. _____

        1. _____

        2. _____

    B. _____

        1. _____

        2. _____

II. _____

    A. _____

The same thoughts can be "grown" organically. An example of this is found on page 95.[4]

For many of us, this method of sketching ideas is closer to how our thoughts naturally grow from one to another. Later, this pictorial outline can be translated into the linear type given previously. This translation could be important if the thoughts have to be written into a report, which by nature is linear (i.e., having one thought after another, one paragraph after another, one chapter after another).

Meditation

Almost every day-to-day activity requires some form of concentration. Without concentration, we could not make a presentation, weld a joint, or write a computer program. In athletics, the optimum state for peak performance is one of "relaxed concentration" or "playing loose." Similarly, meditation is a very effective way to bring yourself into a more focused state of relaxed attention.

The term *meditation* is often regarded with suspicion in the business

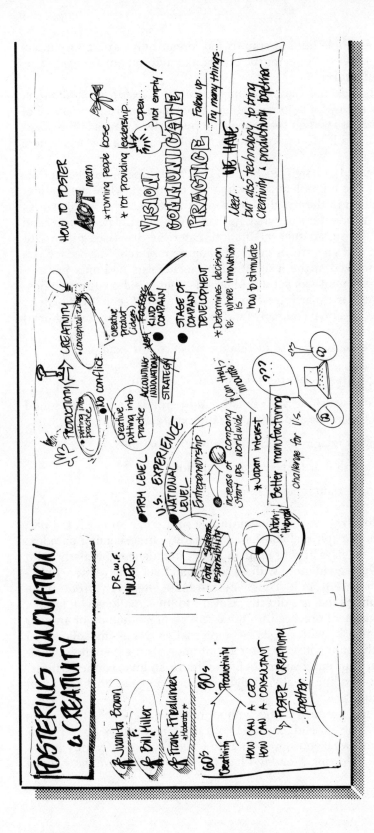

FOSTERING INNOVATION & CREATIVITY

DR. W.F. MILLER...

Juanita Brown
Bill Miller
Frank Freelander *Moderator*

60's
"Creativity"

80's
"Productivity"

HOW CAN A CEO
HOW CAN A CONSULTANT
→ FOSTER CREATIVITY ...Together...

PRODUCTIVITY ⇄ CREATIVITY
• Putting into practice
• Conceptualize
• No conflict

Creative Putting into practice
Creative Product (idea)

• FIRM LEVEL
• ACCOUNTING INNOVATION - BEST PROCESS KIND OF COMPANY STRATEGY
U.S. EXPERIENCE
• NATIONAL LEVEL
Entrepreneurship
increase of Company start ups worldwide
Can Tech? innovate?
* Japan interest
Obtain "Hybrid" Better manufacturing
Challenge for Us.

STAGE OF COMPANY DEVELOPMENT
* Determines decision ie where innovation is
how to stimulate

Total Systems Responsibility

???

HOW TO FOSTER
NOT mean
* turning people lose
* not providing leadership
open
not empty!

VISION
COMMUNICATE
PRACTICE   Follow up
...Try many things...

Idea   WE HAVE
but also technology to bring creativity & productivity together.

world; some people think it sounds too foreign, mystical, and impractical. Perhaps better terms for the corporate environment are *mind clearing* or *focused concentration*.

Meditation is not, as some might think, a losing of awareness. Rather, it is a *heightening* of awareness to include your inner source of creativity. It helps you go below the choppy, surface waves of the ocean (the external world) to explore the depths beneath the surface, where the treasure is.

One talented research biochemist, Dr. Philip Lipetz, received a startling insight into a problem of biochemical pathways while meditating. After experiencing the insight for the first time, he went to his books and found no mention of what he had envisioned. The same vision came again during another meditation. Later, he discussed his insight with an eminent scientist in the field of aging research. The second scientist was amazed at the insight; other biochemists had only just finished an experiment—as yet unpublished—that could confirm the possibility of the particular interaction Dr. Lipetz had "seen." Ultimately, Lipetz started his own company (and later his own venture capital firm) to develop this idea.

There are many types of meditation practice, for example:

- Walking with special attention to your movement

- Sitting and watching your breath, and perhaps hearing the sound "so" on the inhale and "hum" on the exhale

- Watching or imagining a candle flame or a mandala

- Having a vision or fantasy (guided or not)

- Following certain types of music

When you touch your innermost intuitive self, you also touch the part we all share, the spiritual core of all humanity. In a seeming paradox, you can find oneness in your ultimate, inner uniqueness. Similarly, with meditation you can identify with a problem and yet achieve more objectivity and distance from it. Thus, you can more effectively produce the creative solutions that are directly relevant to the situations you face.

With consistent practice through one type of meditation or another you gain familiarity with the states of internal awareness that allow you to listen to the still, soft voice of your intuition. You experience a communion with your inner source of life—with trust, love, connectedness, and wisdom—finding creative insight within "the peace that passes all understanding."

I would not say such things without having experienced the results myself. One time I was dissatisfied with my career progress, so I meditated over the weekend on, "What do I need to do differently in my career?" The following Monday I had the answer, and two days later I found out about a position that I probably would have overlooked had I not asked my intuition for guidance. That began my current career.

I have also successfully begun creative strategy workshops using drawings based on meditation that symbolize important themes for each workshop. And in a nonwork example, in September and October 1982, I was gifted through meditation with the awareness that on the coming Christmas Day I would meet the woman who would become my wife. I even bought her a Christmas present ahead of time. Sure enough, I met her on Christmas afternoon!

With the themes of peace, creativity, and unity being so closely associated with meditation, meditation can be conceived as a possible twenty-four-hour state, as much a part of your life as breathing. In fact, one of the principal techniques for developing the state of meditative concentration is to focus on our breathing. Try the following exercise for a few moments after reading it, even if you have already experienced something similar to this before.

Sit or lie down comfortably with your back straight but relaxed. Slowly close your eyes. Notice your breathing. Is it relatively shallow or deep? Does your breathing start with the upper chest or with the lower diaphragm?

Just notice the way it is and do not try to change it. Let it change of its own accord as this exercise proceeds.

Imagine a warm, golden, glowing ball of light suspended six inches above the crown of your head. Warm, golden light is streaming from this globe down through your head. See and feel this warm, golden light flow into you, gathering up all tension as it passes through your body. *Slowly* let this light wash down through your head . . . into your neck . . . your shoulders . . . your arms . . . your hands . . . your chest . . . your back . . . your abdomen . . . your thighs . . . your calves . . . your feet . . . and out your toes into the earth. Let this wash through you and carry with it all tensions, worries, anxieties, tightness . . . all thoughts and emotions . . . leaving you soothed and relaxed.

Now feel your breath passing through your nostrils. Follow each breath into your lungs, and notice whether the air fills the top or bottom of your lungs first. Notice whether your exhale begins at the top or bottom. Notice the time it takes to inhale and exhale. Just notice, and do not try to make any changes. If changes occur, let them do so of their own accord, from their own wisdom.

Listen and imagine the sound "so" accompanying each inhale, and the sound "hum" accompanying each exhale. Listen to the "so-hum" of each breath cycle. Notice any differences in your state of relaxation and awareness. Continue following your breath in this way for as long as you wish.

If you wish, imagine that the air you breathe is really the warm, golden light entering your body. With the sound of "so," it caresses your whole insides, cleaning and energizing your transparent body and mind. With each exhale and the sound of "hum," the golden light exits, carrying any tensions or thoughts.

Now go ahead and do this exercise for five minutes or longer. Slowly come out of the meditation and reflect on the experience. Make a request of your intuition for an insight, such as, "Please give me an image or symbol of the most important lesson for me in this book." (It may be something covered in a later chapter.) From meditative quietness, your mind can give very rapid responses to questions that would otherwise continue to puzzle you. In large part, it is a matter of trusting these messages.

An Example of Fluency

A fluent combination of intuition and logic can be a powerful tool inspiring people to undertake a meaningful mission and to be open to learning. I once worked with a container company on a business strategy that involved the subject of recycling. We used a combination of linear and intuitive techniques to develop the alternative strategies for introducing a new container.

We initially stated the objective as developing an economically feasible, practical strategy that would meet environmental concerns for the new container. The container had many potential environmental advantages of which the public was generally unaware.

One of the linear techniques was to portray the arenas for strategy alternatives as a 3-D matrix in which we pictured various mixes and matches of issues, strategy components, and stakeholders.

ISSUES	STRATEGY COMPONENTS	STAKEHOLDERS (those with an interest in the final solution)
Recycling	Economic/technological incentives	Consumers
Litter		• traditional middle class
		• achievers
	Education/PR	• societally conscious
		• others
		Industry corporations
		Legislators
		Others

First taking recycling as a key issue, we explored many of the demographic and attitudinal descriptions of the society and then brainstormed strategy alternatives in each of the "strategy component" areas. In some cases we said, "Pretend you are a member of this particular consumer group; what would you want to see as the final strategy of this corporation?" Once we had developed strategy ideas with recycling as the prime issue, we did the same with litter.

A primary component of creative strategizing is to alternate between linear and intuitive techniques. This keeps participants refreshed and allows us to get maximum practical insight. After the 3-D matrix exercise, the participants formed groups of threes—taking the roles of consumer, corporation, or legislator. They were instructed to maintain silence and to use crayons to "negotiate" (by making lines, symbols, figures, etc., on paper) a solution where everyone wins: a portrait of a win-win solution. Although at first unsure of what to draw, they eventually produced a visual statement that we could discuss and determine the qualities the images represented. This became important information to store away, to revisit after more time with more linear techniques, such as a brief version of alternative scenarios.

Near the end of the day I had the participants close their eyes and, with music, led them into a deep relaxation state. Using very general instructions to guide their imaginations, I took them on a tour of a vacation spot of their choosing. After walking on a path awhile (one person, I found out later, was canoeing on a stream), they saw something that caught their attention and examined it closely. Then they attended a party in their honor where different stakeholders came up to congratulate them on the final strategy.

When the exercise was over, we discussed the qualities they discovered in whatever had caught their attention, plus the comments made to them at the party. One person saw a snowflake and fresh snow at a ski resort, and for him this symbolized the new and somewhat fragile territory his company was embarking on. Another person saw a fish he had never seen before, swimming slowly and unafraid in the currents. He saw that the situation was new yet familiar; and when he asked the fish to speak, it advised "move slowly and test the currents, but be unafraid." Other images reflected very similar themes (a good test of the intuitive "rightness" of the group's perception of the optimum solution).

As a result of both the linear and intuitive idea generation, four possible strategy themes emerged for the company, some more aggressive in promoting recycling and some more conservative. A summarization of costs/benefits/risks was drawn up and presented to executive management showing what research would be needed to verify how each strategy alternative might: (a) differentiate the company from its competitors; (b) gather higher market share; (c) improve legislator relations; (d) potentially raise stock values; (e) implement corporate concerns for the environment; and (f) provide the highest win-win for all stakeholders.

IN CLOSING . . .

■ *One of my favorite things is to try to build my mind by thinking about something that really goes beyond, like the universe—Where does it end? What is beyond the universe? You could go on for hours and hours thinking about that, and trying to figure it out strengthens your mind. It's hard to explain. You have to go deep back in your mind to understand what the mind can do. It's unbelievable how far you can go.*

Stan, private in the U.S. Army ◗

■ *When I teach a seminar, there's really a state I get into—sort of a mind-set. You know: you read the audience, you respond to the questions, and you obviously don't have time to think about it on any level. It's a state of being, a state of mind, a being "present."*

Bruce, litigation analysis consultant ◗

Intuitive approaches to creativity can open our minds to our internal wealth of insight and wisdom that we might not otherwise tap into. Training in these methods becomes an important new dimension to using the more linear, or logical, idea-generation techniques.

When we've developed fluency with both linear and intuitive methods, we can often instantly translate our intuitive "leaps of logic" into language, and our ideas come out as verbal ideas. It may happen so automatically, so quickly, that we don't even realize how truly fluent we have become in creativity.

Thus, you can realize a stronger connection between your intuitive answers and your deeper sense of purpose and meaning in life. Your answers—to whatever problems—will begin to feel richer, with more personal intention to make them real in the world.

CHAPTER 7

Transforming Individual Blocks to Creativity

Blocks Schmocks

🖝 *Sometimes I get separated from my work. I was real worried about how a piece was going to be viewed by my peers, and I just took all that totally to heart. It just screwed the whole thing up. It became an object and then it didn't work anymore. I didn't have any rapport with it . . . I totally gave away my power on a creative level.*

It's quite different when working with somebody on a commission. Working under those conditions—creating a commission for somebody and working with their ideas—now that's a really creative process because I have to be able to tune in to those people. I have to pay attention not only on a creative level, but listening and paying attention to what their needs are.

Patricia, artist 🖝

In chapter 4 I made the somewhat bold assertion that we are *always* being creative and it simply depends on *what* we spend our time creating that makes a difference. Einstein's theory of relativity was a more inclusive scientific principle than were Newton's laws of mechanics, without negating Newton's laws. In the same way, this broad notion of creativity is a more inclusive statement for exploring the habits that support or inhibit our inherent creative abilities.

With the broader context firmly in mind, we free ourselves to refocus our creative energies in each moment of our work lives. Our every phone call, every report, every turn of a screwdriver, every meeting, every policy implementation, every strategy formulation can be experienced in terms of our creativity. We all have times, however, when it seems that the ideas and insights just don't come, try as we might. Perhaps the true test of our creativity is finding new ways to transform those "blocked" moments. It is much easier to simply refocus our creativity onto new arenas than to try to go from being "uncreative" to creative.

The chapters in Section III of this book cover blocks to innovation and creativity for which group solutions are appropriate. This chapter deals with transforming these blocks on the individual level.

In chapter 2 you read about a consumer products company's strengths and improvement areas as an organization. When we asked the people in this company to describe their predominant individual blocks to creativity, they responded:

- Accepting conventional wisdom as an appropriate approach

- Lacking time to investigate or elaborate on new ideas

- Seeking only to satisfy the perceived needs of management

- Having tunnel vision, compartmentalizing problems

- Looking for quick yes/no answers

- Fearing rejection of ideas

- Being afraid of making mistakes

- Expecting others to be the creative ones ("I'm not creative" or "It's too hard")

- Being unwilling to question others

- Being unwilling to accept others' input

- Being unwilling to collaborate

In other surveys a few other individual blocks to creativity were added, which together make up the following generalized list of blocks:

1. Emotional

2. Personal/cultural perceptions

3. Social/expressional

4. Preferred method(s) of problem solving

5. Skills/specialization

6. Stress

7. Lack of Imagination (Image-ination)

Looking at such a list, you might have the impression that you could never be creative without a lot of effort. However, there are some powerful principles, similar to the APPEARE process, that can help you develop an internal ability to be creative rather than blocked.

Let's review the principles of transforming blocks before looking at each type in detail. As you then read through the rest of the chapter, you can imagine how you would effortlessly dissolve a block.

The Art of Transforming Blocks

The art of *transforming* our blocks is different from trying to "change" ourselves. To change, we must battle our past and present and somehow make our lives different. In transformation your energy can be devoted to building a new, chosen "reality." Rather than fighting the present situation, you simply acknowledge what's happening and then focus on what you want to replace it.

The basic steps in transforming old habits (especially blocks) are these:

1. Assume that every experience you have can positively stimulate your personal growth.

2. Discover the fundamental truth and the fundamental illusion in the barrier(s) you face.

3. Clearly envision how you want to be (and what you want to be doing and having).

4. Evoke the situation and emotions surrounding the block, holding strongly to your vision and your inner strength.

5. Take action and/or communicate with another person.

6. Allow yourself time. Have patience and honor the seasons of creativity.

Transformation takes place when a new world view (paradigm) replaces an old one.

1. *Assume that every experience you have can positively stimulate your personal growth,* empowering you as never before. (This step is not required for following the subsequent steps, but it sure helps.) Each of us is on a path of personal growth and integration. But sometimes, or often, we make incorrect decisions about life. Indeed, we may feel like we are encountering events in life without an instruction manual on living. But as Robert Louis Stevenson once said, "To be what we are, and to become what we are capable of becoming, is the only end in life."

2. *Discover both the fundamental truth and the fundamental illusion in the barrier(s) you face.* If we are afraid of failure, the fundamental truth may still be that we want to make total use of our talents to make a real contribution. The fundamental illusion may be a belief that we will be worth more or less as a human being if we don't make an impact. One way of discovering the truth is to ask yourself, "What is my highest goal for this situation?" One way of finding the illusion is to write down a fantasy of the worst that can happen, the thing that makes you think "I'd die if this happened!" (which will rarely be the truth).

3. *Clearly envision how you want to be (and what you want to be doing and having).* Use all the senses and emotions in your imagination to visualize the final result. Develop a statement of affirmation stating your goal *in present time.* For example, if you are stuck seeing a problem only one way, your affirmation might be, "I am now able to see this problem in a multitude of ways, opening before me a multitude of solutions." If you begin arguing with yourself ("Oh, I could never be that way!"), take two pieces of paper and write the affirmation on one side, the "answering argument" on the other, then the affirmation again, the new answering argument, the affirmation, the argument, and so on. You will eventually feel a shift, a sense that you can be the way you would want to be. Very importantly, you can also call on an inner strength to empower you to fulfill your vision of yourself.

4. *Evoke the situation and emotions surrounding the block, holding strongly to your vision and your inner strength.* Emotions such as sadness, pain, and anger are signals to make a shift, to be more "on the mark" rather than to "sin" (which literally means to "miss the mark"). A barrier is simply a feeling or situation to which you have given over your own personal power, leaving you feeling powerless and afraid of it. By preparing your vision of yourself, taking on your inner (higher) strength, and then commanding the barrier to come before you, you take your power back. The energy tied up in the former battle becomes available to you for your creativity: the "demon" you faced becomes your ally.*

---

* This method is present in all the mythic journeys. In the Odyssey or the travels to Shambala, for example, there are occasions when the adventurer comes to the impassable river (or other obstacle), guarded by a demon. The instructions are clear: prepare yourself meditatively and gather strength; identify with a Power (Divine) so its energies merge in you; then call forth the demon, battling until victorious. Having subdued the demon, it then becomes an ally in your getting across the impassable obstacle. There truly is some basis for saying that each situation is presented to us for our unfoldment into more power, love, and wisdom as we live and work.

5. *Take action and/or communicate with another person.* In speaking or in taking other action, your newfound power takes root. You make it more real in your own and others' eyes. At work this can mean demonstrating your new creative insights or your new ability to follow through on your ideas. Developing a support network of people you trust can help you explore and gather your new power and insights in a safe environment.

   a. A useful principle in taking action is to draw the distinction between handling a situation and handling our emotions in that situation. They are entirely different matters, though it is typical to assume that the situation *causes* our feelings, and that the only way to get rid of uncomfortable feelings is to change the situation. Actually, the more we give power over our feelings to the people and situations we encounter, the less able we are to respond to them creatively and effectively.

   b. As before, Japanese self-defense teaches an important lesson about life: if we emotionally reject the fact that we are being attacked—with either a fearful cringe response or an angry urge to kill or injure—we become *more vulnerable* and *less centered* in mind and body for taking successful action. By giving power over your emotions to the situation or opponent, you become less able to handle either the emotions or the situation.

6. *Allow yourself time. Have patience and honor the seasons of creativity.* None of us can keep running without rest or nutrition. Similarly, none of us can be constantly "on," with full-speed creativity and innovation. Transformation sometimes happens slowly, like winter to summer, or ice-age sea to desert. These seasons and cycles need to be honored rather than impatiently judged.

There are two workable approaches to honoring the pace of transformation. The first is to make an affirmation that is more than just "a good idea" or "something I ought to do." The commitment must be intuitively and emotionally based.

The second is what I call "the hot-stove method." Whenever we touch a hot stove and experience pain, we immediately and spontaneously pull our hand away: we naturally do what is in the best interest of our well-being. There may be many situations in your life that hurt your well-being in subtle ways, and you are numb to the hurt. As you become more conscious of the damage to your well-being, you will find yourself naturally and spontaneously making changes in your life to "pull your hand away."

The hot-stove method describes the way I have significantly changed some work habits. I used to automatically presume nonsupport for a new idea I might have. Immediately I would be in a "fight-for-what-you-want" posture—and then feel self-righteous when the resistance I expected materialized. Finally I realized I was creating much (or all) of that resistance by my attitude. It was painful and difficult to face this—unnumbing to how I was "burning" myself. Slowly, over the course of a year, I found myself changing as I watched myself at first fight and then let go of the need to do so. I held the vision that "My work environment supports my ideas. Even when the ideas get changed, they contribute to the highest good of all."

Transforming Different Types of Blocks

All of the seven types of blocks—emotional, personal/cultural perceptions, social, problem-solving methods, skills/specialization, stress, and lack of "image-ination"—can be transformed for both personal growth and improved organizational productivity. In some cases, time, practice, and/or training may be needed. Other changes may be achieved relatively quickly. Understanding more about each block can help clarify what you want to achieve.

Emotional

Some of the emotional patterns that inhibit creativity and innovation include the following:

- Guilt, depression, or anger—indulging in negative thoughts about the *past:* "Oh, if only I hadn't done that . . ." "If only he/she hadn't done that!"

- Worry and anxiety—indulging in negative thoughts about the *future* ("Oh, what will tomorrow bring?")

- Comfort with the status quo ("Why change? If it's not broken, don't fix it.")

- Boredom ("This doesn't interest me.")

- Excess energy ("I can't sleep on it; my mind's too busy.")

- Intolerance of ambiguity ("I need an answer *now!*")

- Fear of success ("Do I really deserve success? What will people expect afterwards?")

- Fear of failure ("What if I fail? I'll look foolish and be vulnerable.")

- Need for perfection ("If I can't do it perfectly, I won't try it at all.")

- Excessive zeal ("I'm a great scientist. Of course this invention is great. Everyone will want one.")

- Disinterest ("What's the use? Management won't go for it anyway.")

- Impatience and frustration ("Why isn't this working out?!")

- Seriousness ("Let's stop playing around. We have some serious creating to do.")

Creativity at work includes how we create our immediate emotional world. Have you ever noticed how much better you feel around those who have a cheery disposition than around those who don't? Situations don't cause our feelings; our perceptions do. When our perception is that "the situation causes my feelings," then we give away our inherent power over internal experience. We can reverse the power flow by bringing our own cheeriness to the moment and letting others respond to our mood (rather than responding to theirs).

One example of emotional blocks hurting innovation is the "not invented here" (NIH) syndrome. We sometimes disregard good ideas we don't think of personally. Sometimes it makes good business sense to do so (proprietary technology development, perhaps), but the NIH decision is often made on a macho ethic: "Truly competent businesspeople don't rely on others to fulfill their responsibilities." NIH people might insecurely think, "I'm getting paid to come up with ideas and make them happen. If I use this idea, it shows that I can't do my job."

There are several guidelines to keep in mind.

1. Prepare a vision of feeling differently. A good affirmation to keep in mind is, "I am affected only by my thoughts."

2. Be willing to use that vision to become bigger than the inhibiting emotions. Exchange thoughts of fear for thoughts of competence, confidence, love, and so forth.

3. Watch out for suppressing feelings in a bliss of positive thinking. Find the truth and illusion in them and evoke them to consciousness. (You might sometimes choose to remain in a work situation in which you experience discomfort until it no longer "hits your buttons.")

4. Nurture alternative inner experiences ("inner" creativity). Then, over time, you can take back your power over how you feel (and subsequently act).

And again, watch out for believing that the situations you face are the causes of your negative feelings. In such cases you might feel righteously justified in hanging onto your (prejudicial) thoughts and emotions. There is a well-known story about a psychologist studying the behavior of rats in a maze.[1] The psychologist had a box with five tunnels leading away from an open area. He put some cheese at the end of the second tunnel; when he put the rat into the open area, the rat ran down the first tunnel, found no cheese, ran down the second tunnel, found the cheese and ate it, and then went down the other tunnels to see if there was more cheese.

The psychologist always placed the cheese in the second tunnel before placing the rat in the open area. He wanted to see how long it would take for the rat to figure out that the cheese would be in the second tunnel and head directly down that tunnel. After a while, the rat didn't even bother going down the other tunnels once he had gotten the cheese in the second tunnel.

One day the psychologist switched tunnels and put the cheese down the fourth tunnel. When placed in the open area, the rat went down the second tunnel, found no cheese, came back to the open area, went back down the second tunnel, found no cheese, came back out, went down the second tunnel, found no cheese, came back out, and so on. Eventually, the rat started exploring other tunnels until it found the cheese.

This story shows that there is a very important difference between that rat and many people. The rat eventually explored the other tunnels when there was no cheese in the second tunnel. Many people, however, continue going down the "second tunnel" of their lives believing that at least they are in the "right" tunnel.

Personal and Cultural Perceptions

When I was in graduate school, one faculty member was a woman from Thailand who had come to the United States to study psychology with Abraham Maslow. She told of her first months in the States when she had observed people scurrying about, accomplishing many things but apparently with no inner peace. This puzzled her until she realized that many Americans seemed to have been raised with the notion that, "You're not worth anything until you prove yourself. And once you have proven yourself (to society/family/community), you must continue to prove yourself over and over." Her upbringing had taught her, "You are valuable just by being part of the universe. Your value cannot be gained or lost. And the art of life is living in a way that reflects that fact about everyone."

This story continues to speak to me when I feel insecure or rebellious. I ask myself, "What am I trying to prove to myself or others? Do I really *need* to? Can I still act but without these feelings driving me?"

When we are young, we come to decisions about life that may be perfectly appropriate at the time but that may or may not fit the circumstances of later life. These decisions become fundamental life principles that shape the course of our perceptions, thoughts, feelings, actions, and visions. These fundamental principles are not *what* we think and feel; rather, they are the *context* for what we think and feel. They are the lens through which we consciously and unconsciously live our lives. They are the box we live in.

One example of a fundamental life principle is "I must get approval from others to feel okay." Then our entire life is lived in the box of trying to get other people's approval at any cost. Other examples include "Be careful," "You've got to win," "Always play it safe," "You've got to make it on your own," "You've got to be right," and "Be nice."

These fundamental life principles result in blocks to creativity expressed as: accepting conventional wisdom ("The experts know"); tunnel vision ("I'll define the problem in ways that I know how to answer"); and social blocks, such as those discussed later in this chapter ("I'll only do things I know will be accepted by others").

Our culture has principles that can also block creativity, including:[2] "Fantasy and reflection are a waste of time." "Playfulness is for children only." "Reason and logic are good; feeling and intuition are bad." "Tradition is preferable to change."

From my experience, there are five categories of fundamental attitudes that affect creativity. The five are:

1. *Leader/Victim.* Do you generally feel that you are directing your life or that you are the victim of outside forces? Do you have a sense of personal power?

2. *Confidence and Trust.* Do you tend to see the world as friendly or antagonistic? Do you have confidence in yourself (and/or a higher power) for having your life "turn out" okay?

3. *Talent/Contribution.* Do you feel you are exercising your talents and having a positive impact on the world around you?

4. *Strength of Expectations.* Do you feel "addicted" to having life turn out according to your expectations? Do you problem solve or do you blame when your goals aren't accomplished?

5. *Personal Value.* Do you feel you have to prove yourself continually to "earn" self-esteem and the esteem of others, or do you believe in an inherent personal worth of each person?

Once you become aware of your fundamental life principles and recognize their power to shape your interactions and communications, you can intentionally give new purpose and direction to your life and your creativity. If you merely extend yourself from the past with a few "changes," there is no transformation, only more of the same.

Social

The social blocks are emotional barriers or fundamental life principles that concern our relationships with our managers, peers, friends, and others. You may recognize some of these in yourself or others.

- Lack of confidence in personal creativity ("Other people are creative, not me." "It's too hard." "This is too ambiguous.")

- Fear of rejection ("What will they think of me if I'm not right?" "I'll look foolish." "Play looks too irresponsible.")

- Need to please management ("I must be cautious and conform. My job and life are on the line." "Follow the rules.")

- Unwillingness to collaborate ("Please, I'd rather do it myself." "That's not my area.")

- Fear of confrontation ("I can't hold my own with this person." "Harmony at all costs.")

Social blocks are acted out either as the inability to express or the distorted expression of an inner need. Expressional blocks may be based in inadequate skills for expressing an idea—whether in words, writing, pictures, music, or whatever. Training in various communication skills can often be the most effective means of transforming these blocks.

Distortions occur when we somehow decide that we *can't* get an important need met or that there is only one particular way to meet it. At work you probably observe actions that *don't* support people's talents or dignity. These distortions hide a positive need that is being expressed negatively. For example, when we show off, the positive need is to feel socially connected, but we have decided it's only possible to feel connected by showing off (or that it's not possible at all to feel connected). Or when we are selfish, the positive need is to feel our uniqueness. When we are overly sacrificial (to the detriment of also getting our own needs met), the positive need is to contribute.

A way out of these distortion social blocks is to discover this positive need and creatively find new ways to express it with others.

Preferred Problem-solving Method(s)

We may have favorite problem-solving modes, and these may keep us from working on a problem in the easiest mode. Take for example this story from James Adams's *Conceptual Blockbusting*:[3]

> One morning, exactly at sunrise, a Buddhist monk began to climb a tall mountain. A narrow path, no more than a foot or two wide, spiraled around the mountain to a glittering temple at the summit. The monk ascended at varying rates of speed, stopping many times along the way to rest and eat dried fruit he carried with him. He reached the temple shortly before sunset.

After several days of fasting and meditation he began his
journey back along the same path, starting at sunrise and
again walking at variable speeds with many pauses along
the way. His average speed descending was, of course,
greater than his average climbing speed. Prove that there is
a spot along the path that the monk will occupy on both
trips at precisely the same time of day.

Did you solve the puzzle? More importantly, for our purposes, can you remember *what thinking processes* you used in working on the puzzle? Did you verbalize? Did you use imagery? Mathematics? Did you consciously try different strategies on the problem?

If you happened to choose visual imagery as the method of thinking to apply to this problem, you probably solved it. If you chose verbalization, you probably did not.

A simple way of solving the puzzle is to visualize the upward journey of the monk superimposed on the downward journey; if it helps, imagine two monks, one leaving the top and the other the bottom, starting at sunrise. It should be apparent that at some time and at the same point on the path, the "two monks" will meet.

Some of us like to use our intuition without previous investigation of details and data. Other people like to find the mathematical formula that expresses probabilities and proceed from there. Others like to start with verbal descriptions of the problem to be solved, while still others like more pictorial problem statements.

The previous two chapters gave many different methods for generating ideas in linear and intuitive ways. A third way of thinking that's important to creativity is *judicial*. The judicial is used for evaluating and testing the innovative ideas and efforts against the desired, valued results.

Each way of thinking is important at its own time. For example, you might apply judicial thinking during idea generation by mistake, with idea killers such as: "Let's send it to committee." "It isn't in the budget." "It's against company policy (or strategy)." "Has anyone else ever tried it?" "If it's not broken, don't fix it."

When looking for innovative solutions, sometimes our usual "tunnels" of problem solving just don't have the cheese. By staying in the most appropriate style of thinking for the stage of the creative process, you keep your perceptions open to the most relevant information. As you develop flexibility in trying different approaches to solving problems, you free yourself to give creative service to your organization and the world.

Skills and Specialization

Sometimes the specialist knows too much. In one workshop on new metal materials, we had to invite some nonmetallurgists to help generate the new ideas; the metallurgists knew too much about what you *couldn't* do with certain metals.

Specializing can be both freeing and confining. It is freeing in that specialization enables us to express ourselves in unique ways. A dancer who works out eight hours a day becomes free to move his or her body expressively and uniquely; without the practice and discipline, the freedom is lost.

Specializing can also be confining by dominating our time and restricting us from developing other viewpoints and new talents. Painters know that working in new mediums—from oils to chalks—not only gives them new modes of expression but also new insights into what they can do in their more familiar mediums. Sculptors, like Rodin, often have also painted to keep their creativity free to grow.

Stress

There are three experiences for which we popularly use the word "stress" in our culture:

1. Stress experienced as stimulation from outside of us—this can be either constructive or destructive for us

2. Stress experienced as an affliction of tension, anxiety, and physical impairment

3. Stress experienced as the internal urge for growth and expression

Stress as stimulation is something to manage. Stress as affliction is something to reduce. Stress as the urge for growth and expression is something to be promoted (in the rhythm of stress, rest, and nutrition).

Stress is fundamentally this latter urge to express our aliveness, to put our human awareness into action. Stress as stimulation is a subset of this urge—the part that reacts to perceived threats with a flight-or-fight response. And stress as affliction is a subset of stimulation stress, reflecting a mismanagement of the stimulation.

Creativity and innovation can be spurred, at least in the short term, by any of these three experiences of stress. But a constant push to be working at 110 percent—with continuous "urgent" projects, for example—can promote an excess of stress as outside stimulation, leading to afflictive stress and low-quality output.

Afflictive stress arises when our creativity in life is misplaced, covered, or forgotten. Our creativity is generally enhanced if stress comes from the power of our own urge to grow, taking on challenges to be all we can be. As Maya Pines states in discussing her research on stress and psychological hardiness:[4]

> Stress-resistant people . . . have a specific set of attitudes
> towards life—an openness to change, a feeling of involve-

ment in whatever they're doing, and a sense of control over their lives.

Internal stress arises from moving away from your zones of dependable strengths. Sometimes this stretch is healthy, like "pushing" in physical exercise to get stronger.

With this expanded notion of the nature of stress, you no longer face a question of "responding" to stress or "managing" your stressful situations and internal responses. You can shift your attention to creating a new harmony between your internal nature and external world.

This is the harmony seen in athletes in peak performance. For example, John Brodie, former pro-football quarterback, once said:[5]

> *Sometimes I let the ball fly before Gene [Washington] has made his final move, without a pass route exactly, it's sort of intuition and communication . . . You can get into another order of reality when you're playing that doesn't fit the grids and coordinates most people lay across life.*

In this "flow" state, athletes are alert and active yet loose and fluid. They integrate awareness and action without "talking" their way through the steps of performing.

For further ideas on overcoming stress blocks to creativity, there are numerous books and courses giving specific details and counseling on ways to reduce afflictive stress, manage stimulating stress, and actualize internally generated stress for growth.

Lack of Imagination (Image-ination)

Sampson Raphaelson, screenwriter of the first talking movie and famous playwright of the 1930s and 1940s, defined imagination as "the capacity to see what is there." The "lack of image-ination" is usually merely a lack of conscious practice of our imaginative abilities.

For example, we need imagination to experience fear, because fear is an emotional reaction to our imagining an unwanted future situation. Even when facing a concrete, impending danger, fear arises in conjunction with an imagined, unwanted outcome of the situation perhaps to occur only a few seconds later. Fear, therefore, is successful imagination, whether or not it is appropriate or misdirected.

More often, we may simply be out of practice in consciously eliciting images with all our senses—sight, hearing, taste, smell, and touch. As an exercise, please imagine the following:

- A yellow daisy

- The voice of Frank Sinatra

- The smell of bacon

- The smoothness of glass

- The tartness of a lemon

- The sound of a trumpet

- The sweetness of pure honey

- The coldness of an ice cube

- The smell of fresh-cut grass

You may have noticed yourself being more attuned to one type of sensory imagery than another. This is common. You might even have noticed that your body reacted physically to some of the images, such as the cold ice or tart lemon.

You can also manipulate your imagination. Now, please imagine the following:

- A horse jumping on a huge trampoline

- A large glass of milk being spilled on a table and running onto the floor

- The table with milk on it moving its legs to also "run" across the floor

Your imagination can be improved by having fun with images such as these.

IN CLOSING . . .

▟ *"About two years ago I had absolutely the worst day of my life. Everything went wrong . . . Projects didn't work; people didn't respond the way I wanted; there was a mess and the overload was getting greater; everything that happened all through the day failed. And I was real depressed at the end of the day. I went off to a coffee shop—a high-performance place for me—and I wrote on the top of the piece of paper 'Lessons Learned and Re-Learned.' I went into some kind of an altered state and came out of it hours later with twenty-seven lessons learned, listed on about seven pages of paper . . . just constantly writing things and totally immersed. At the end of that I started writing a 'Lessons Learned' journal every day for two years.*

At the end of that (first) session, I felt totally contented, joyful, and successful and realized that I had just had one of the best days of my life. And ever since then, it was like a total attitude shift on the dimension of what is failure, what are problems. The attitude . . . was to enjoy the lesson and not get caught up in the self-blaming and judging criticism, defeatism and all that stuff. Now, very often, at the moment of something not working out, I say, 'Ah, it's going to be great this evening when I write the lessons learned and can note this one—and actually feel good right at the moment and not have it weigh on me.

So in the context of creativity this attitude is: Every moment is a creative wellspring for growth—including, and maybe especially, the ones that don't work out quite the way you thought. (With) the ones that work out the way you thought, there's probably not that much to learn . . . The ones that are surprising are the ones that can be milked."

Ron, marketing consultant and professional speaker ◗

I'm fond of an expression, "Acknowledge the dark, dwell on the light. Abundance and love for all." Our creativity is developed by recognizing the way things are, including the things we consider dark or negative, and then leaping forward to envision what we want to create. We do not have to try to change anything about ourselves. We just have to focus on creating whatever way of being our inner wisdom advises.

When you reach barriers in your path, you can prepare and evoke a newfound strength from within. Calling forth your fears, angers, or external obstacles, you can turn the negative energy into your ally and do what you once thought was impossible. This lesson is truly applicable to facing the challenges of being creative in your work world. As you successfully meet your challenges, you will find that each situation can be a source of more power, love, and wisdom.

As you have seen, the balance that most promotes our creative self-expression is the harmony of surrendered inner peace and external action. It is difficult to develop this state if we perceive ourselves to be internally split, fighting with our blocks in a tense struggle to be creative.

Focus on some habits you might wish to nurture in yourself and plant the seeds of an affirmed vision. As with any growth, be patient. It does no good to dig up a plant every day to see if the roots are growing yet.

As the Chinese poet Lao Tsu has said,[6]

> *In the universe, the difficult things are done
> as if they are easy.*
>
> *In the universe, great acts are made up of
> small deeds.*
>
> *The wise person does not attempt anything very big,
> And thus achieves greatness.*

And, as a close friend of mine says, "There is no such thing as failure . . . only feedback."

PART III

Groups Working Creatively

"Am I to understand that my proposal is greeted with some skepticism?"

> *Leadership . . . is a role assumed by an individual . . . (It) is not necessarily related to a position within the company.*
>
> —*Colby H. Chandler, Chairman & CEO Eastman Kodak Company*

So far we have been focusing on you, the creative individual. Now comes the question, How do you take your creative actions and potential and make them work within the group for the benefit of the whole organization (as well as yourself)?

Think of this question in terms of your taking a journey. You and others are the drivers supplying both the intention to travel and the spark to ignite the engine. But you still need a destination, a vehicle, an engine, and fuel. The destination comes from the organization's sense of purpose and vision. The organization itself is the vehicle. The steps of the group innovation process constitute the engine. External events surrounding the organization—social, market, political, competitive, and so on—supply the fuel (the "externally supplied motivation").

Pretend you are taking the journey by boat. In some organizations it may seem like you're on a passenger ship with, say, 3,000 people, each using his/her own 100-horsepower outboard motor (with extended propeller shaft), who may or may not be trying to propel the liner in the same direction. Other organizations seem to have 3,000 people purposefully heading in a single direction, with a single 300,000-horsepower engine and a single rudder. Which is closer to being like your organization?

In chapters 8 and 9 you are offered ideas about how you work with the other drivers. In chapter 10 the discussion focuses on the purpose/vision and how it fits strategically with the external environment. In chapter 11 we cover "institutionalizing" innovation and the creative process. Exploring your participation in managing the transition process in organizational growth is the focus in chapter 12.

CHAPTER 8

Establishing Your Role in Creative Groups

Collaboration Plus

🖊 *I was hired essentially to promote anything from a big event—like a "walk" for people over sixty years of age—to something like Christmas camp for seniors. Before I got the job nine months ago, I was on this committee as a volunteer. . . .*

The committee wasn't the ideal situation. With so many tasks to get the Walk going, most of our meetings were very task-oriented. I facilitated the meetings but got caught up in getting things done and creating time lines. There would have been more creativity for me had we shared more about our differences . . . We never did. We generally felt "Let's get it done. Just got to get it done." The more process-oriented way is more involving and more input comes through up front. Fortunately, the Walk came off successfully because we were so adamant, I think.

One of the creative aspects of this project was the mix of people. By osmosis, we drew together four or five people that were very different, very unique and talented, and in very different areas. We complemented one another.

Claire, associate program director for a social service agency 🖊

To take an idea and make it useful and appreciated in the world we need others, and others need us. If we see ourselves as operating in more isolated fashion than we would like, we *can* develop ourselves to be more than a one-man band. There's much we can do to transform territorial battles, to get a project team functioning smoothly, and to develop synergy with different people's talents. This chapter and the next can help you enhance your CREATIVE climate for innovation by focusing on Collaboration, Communion, and Communication in being creative.

When you've been a part of a creative work group, what differences have you noticed? Others answer that question in a variety of ways:

- More enthusiastic interactions
- More risk tolerance
- Less routine—the *process* of work is more invigorating
- Better tools for day-to-day "grunt" work
- Better proposals
- Knowing when *not* to be creative
- More creative delegation

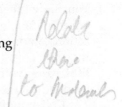

How can creative groups be developed? Fundamentally, it takes skill in promoting a *CREATIVE climate for innovation* in general and in developing *collaborative working relationships* in particular. Of the other seven issues of the CREATIVE climate, of particular concern are the shared vision and the open contact within your competitive environment; these issues are covered in chapter 10.

Collaborative working relationships start with individual responsibility. Every person in an organization has a concrete role in making the organization creatively adaptive to the needs of its customers, stakeholders, and society. Whether you are a white-, blue-, gold-, or pink-collar worker, whether you are management or union, whether you are in marketing, purchasing, engineering, manufacturing, or central staff, you have a role to play. How can you help establish a creative work group?

1. Take responsibility. Make your job and your group positively productive and innovative.

2. Seek out information. Learn about the external environment that is prompting innovations from your organization.

3. Identify administrative obstacles. Seek to find remedies.

4. Promote innovative personnel policies. Institutionalize and encourage individual and group innovation.

5. Participate in and guide organizational change.

6. Ask high-caliber questions. Purposefully stimulate innovation.

7. Apply the organizational vision. Implement the role for innovation.

8. Give and get feedback. Institute a means for evaluating ideas.

Collaborative Working Relationships

Your individual responsibility for innovation is leveraged by how well you can develop synergetic, collaborative relationships. Creative interactions among people can't be forced, at least not for long. However, they don't "just happen" either. Just as a rose bush produces more beautiful roses with selective "structuring" (pruning), innovative groups work best when they pay attention to certain guidelines and structures.

Collaborative working relationships develop when we pay attention to six factors:

1. Informal innovation roles

2. Informal alliances

3. Intergroup relations

4. Leadership and delegation styles

5. Task-oriented teamwork

6. Alignment and attunement

Informal Innovation Roles

Individuals can play many parts in helping a group produce a creative output. It is usually not enough to have a great idea, boundless energy, and a committed "champion." Each of the roles may be the critical ingredient for the group's success.

There are eight informal innovation roles:

1. *The Product Champions.* These people are the energy behind ideas. They will push and shove, if necessary, to try and get their favorite ideas over the organizational hurdles, or they will operate secretively. They are the great believers in their ideas and are willing to take risks to get their ideas implemented. Sometimes their ideas are actually invented by someone else, but product champions may become excited and take them on as their own. They go beyond being the "lonely genius" to use their knowledge of company operations and the market to produce their pet innovations. More often than not, when there seem to be insufficient ideas, people probably are not aggressively championing ideas, either their own or others'.

2. *The Sponsors.* These people support and oversee the progress of the

product champions. Because the product champions are sometimes likely to offend and cause waves in the organization from their zeal, the sponsor buffers the champions from unnecessary interference. This lets the product champions concentrate on the ideas rather than the politics, which may be their weak point. They may also link people in informal alliances (to be discussed) to make better use of complementary strengths.

3. *The Inventors.* These people are the original creators of an idea, a concept, a possibility. They may also be the ones to carry the concept forward as the product champion. In many cases, however, they prefer to stay "at their bench" or in their comfortable research or staff routine. They have mixed feelings if they let someone else become product champion over "their baby," but they may do it anyway.

4. *The Project Managers.* These people are the stabilizers, the ones who know how to manage projects within the triple constraints of performance, time, and budget. They are the touchstone for the product champion to make the idea really work within the guidelines set up by the organization.

5. *The Coaches.* These people give guidance and assist the development of less experienced personnel (a mentor or big brother/sister role). Their ability to provide a role model and to communicate empathy, optimism, persistence, and a critical but trustful attitude can be invaluable to the innovation process.

6. *The Gatekeepers.* These people—and there need to be a number of them—monitor technological, social, political, market, and other emerging external trends and communicate them widely to appropriate people in the organization. This information is most often the stimulus for coming up with the innovation in the first place, as a proactive or reactive response to this input.

7. *The Internal Monitors.* Monitors also are needed to review the ongoing ideas and creative climate of the internal organization. This internal monitoring can be critical to circulating information and decisions and to determining how groups within the organization can respond creatively.

8. *The Facilitators.* These people have a high degree of skill in eliciting new ideas from groups of people and fostering more collaborative teamwork among them. They act as the neutral catalysts and change agents for more rapid, productive interactions. (See chapter 10 for a discussion of a facilitator's skills.)

These informal innovation roles can be remembered using the first letter of each word in the sentence,

For Superior Innovation Projects, Promote
Individual/Group Celebrations!

F = Facilitator; S = Sponsor; I = Inventor; P =
Product champion; P = Project manager; I = Internal
monitor; G = Gatekeeper; and C = Coach.

Notice that this division of roles for the creative process is different from the traditional division between line authority and job expertise. In one case, a top manager may have the internal-monitor role for stimulating a project while the most junior staff member is the product champion. In another case, the monitor, inventor, and product champion may be the same person, and the sponsor might be someone outside his or her line organization.

Yet each role can also be formally established and rewarded. In one major corporation, for example, a vice-president officially appointed gatekeepers in letters by stating:

> In every large organization there are a few individuals who have a special talent for knowing what is going on in a particular technical field and in keeping others informed of these advances. These have come to be known in management literature as "Gatekeepers." We have begun a program to identify and recognize these individuals. . . .
>
> . . . I would like you to write a two- or three-page memo on recent activities and advances in your field of interest. This should not be a primer, but should be addressed to someone already familiar with your area. During the year, as any advances are made and come to your attention, I would appreciate a short note outlining what has happened and what effects this will have on (the company) as you know it.
>
> Get acquainted with those others in the Corporation in your field and include them on your distribution list of documents you generate for your own divisional use.

In some cases, different people must play different roles depending on the needs of the business. If you are producing a new product for a new market, you probably need an "intrapreneur" (product champion) heading it. If you need someone to effectively work with existing products or existing markets, you usually need a different type of leader, more of a "stabilizing manager" (a la the project manager).

During the life of a work group or the course of your career, you may be required to fill all of these roles at one time or another, especially if you are a manager. The skills and interest for fulfilling these informal responsibilities do not always, or perhaps rarely, match up with stated job-performance requirements. Therefore, this may be an important area for human resource development programs in your organization.

Each informal innovation role is important, even critical, to the overall creativity and effectiveness of your organization. Although the sponsor and gatekeeper, for example, may not be viewed as the creative ones in your group, they may be just as important to your group's creativity as the inventor or product champion. With people occupying each role and working together, the creative process can free each person to contribute in his or her best way.

Informal Alliances

You don't have to be a creative superperson. We all have our strengths and weaknesses. To take advantage of different strengths, we can pair up with other people to form SPIRITED alliances. Thus, we can become co-SPIRITEDs (second cousins to "conspirators," who literally are "two who breathe together").

For example, if you are good at generating ideas but have a hard time identifying the one(s) the *others* will most respond to, you can seek out someone who is good at that. Or if you tend to get lost in the details of facts and figures, you can partner with someone who is good at organizing information.

If you are especially good at planning and following budgets, you can help the inventive "idea people" who aren't good at project management. Or if you know the organizational ropes well, you can aid the less politically astute.

Another consideration to keep in mind when we are partnering with people or groups is our different *styles* of creativity. J. M. Kirton has identified two different styles, called "adaptors" and "innovators" (or "originators"). Each may have the same *level* of creativity—and can equally claim the title "creative"—but they exercise their creativity in very different ways.

"Adaptors" like to work within a given paradigm or way of seeing and resolving problems. They are best at taking what's given and working with it in new ways. They can be methodical, disciplined, and dependable. They seem impervious to boredom, able to maintain high accuracy in detailed work. They challenge rules cautiously. They can also be vulnerable to social pressure and can act as an authority within a given structure.

"Originators" like to develop new paradigms and approach tasks from unexpected angles. They like to challenge assumptions and discover problems, perhaps more than solving them. They can maintain detailed work only for a short period and quickly delegate routine work. They often challenge rules and like to take charge in unstructured situations.

Adaptors are like artists who paint landscapes or people as the image is presented to them. Originators are like artists who paint abstract or impressionistic images produced from their mind's eye.

When adaptors collaborate with originators, the adaptors supply stability in the partnership, maintain cohesion and cooperation in the group, and provide a safe base for the innovator to take risks. At the same time, the originators supply task orientation as a break with the past and provide the impetus to overcome organizational inertia. Both need to be consciously aware of the value that lies in their differences. Otherwise, those differences become walls between them.

There are all sorts of times when co-SPIRITEDs are essential to the *group* being innovative and productive. One of the most important talents in promoting group creativity is our ability to identify, foster, and form these bridges. Co-SPIRITEDs typically consist of two or three people, but we can combine individuals, groups, functional specialties, and so on to the best advantage of all. They may be formally recognized and promoted by managers, even when instituting new project teams.

To stretch the concept of co-SPIRITEDs to an extreme, a peak-performing organization can be seen as a formal set of such co-SPIRITEDs. Take engineering and manufacturing, for example. Engineers are talented at making one-of-a-kind solutions to a market need at somewhat great cost. Manufacturers are good at making large multiples of that product at low cost. There is a creative tension in mixing their goals and capabilities: one-of-a-kind products at great cost versus many-of-a-kind products at low cost. Depending on the synergy, this can provide creative stimulus or territorial conflict.

Co-SPIRITEDs can be initiated by anyone. But as with any organizational process, their effectiveness depends on putting together the right resources, communicating a sense of direction and vision, and providing for the right type of feedback. In any case, you as an individual can seek to link yourself with others in such Co-SPIRITEDs, recognizing your strengths and theirs.

Intergroup Relations

When you have an idea you want to actualize, you may need help from many departments—accounting, legal, market research, R&D, and so on. The key to your success may be how well you can "bootleg"—develop allies in other departments. This is a measure of your ability to have an impact on your organization.

In building these relationships across departments—and thus promoting the creative climate for innovation—how well do you encourage the following?

- Sharing of information for both departments to make decisions

- Joint planning and commitment to the project's success

- Equal understanding of the needs driving the project

- A win-win relationship of managerial power and influence

- Visible management commitment of necessary resources

- Sharing knowledge about the other's expertise

- Giving permission to challenge each other

- Avoiding situations where "my need is your problem"

- Consensus implementation of plans, milestones, and feedback systems

- Sharing the glory as well as the effort

To improve the quality of intergroup relations on a large scale, many organizations are making their *management practices and organizational structures* themselves a target for innovative thinking. Chapter 11 has more to say about this.

In other cases, when a fast response to the marketplace is needed, some organizations set up alternative structures (internal venture teams, headed by "intrapreneurs"). The Signode Industries, a $700 million Midwestern manufacturing company, has been forming a series of new venture teams to identify new business opportunities outside the company's normal product areas. Its first team consisted of six senior managers from six different departments—marketing, engineering, manufacturing, and so on. The team's initial task was to investigate the new industry area and recommend specific business opportunities within six months.

One of the interesting aspects of this venture team was that it was purposely established without a leader. The six people were brought together as peers, and they decided to manage the group as coequals. The group's sponsor was the director of corporate development, who wasn't a member of the team itself. The team members each had potential roles as inventors, product champions, monitors, and project managers.

By choosing not to have a leader, they forsook the security of having someone around who would decide what to do next, assign tasks, and step in to resolve differences. The fact that they took such a risk says quite a bit about these managers; their careers could have been affected by the success or failure of this experimental team.

Through a demanding series of over 2,000 personal interviews, they gathered opinions from prospective users and customers in the proposed industry on what new products were wanted. By becoming very aware of the current market situation, they took the first step of the APPEARE process. However, they sometimes felt they weren't being as innovative as they would have liked.

Near the end of their six months of identifying new opportunities, I conducted a two-day "consolidation" session with them. We clustered the opportunities they had identified and shaped them into "businesses" rather than simply a set of related ideas.

Regarding the *content* of their study, their recommendations were ultimately very well received by their company officers and board of directors, and the resulting new venture is now already in the marketplace and doing well. Looking back at the *process* of working together over the six months, team members identified three norms that had helped them work collaboratively:

1. Each person could pursue a potential opportunity even if the others weren't enthused. Each could be a "product champion."

2. Periodically a consultant was brought in to help focus the group's efforts. This helped to synthesize their findings and resolve conflicts of priorities among the group members.

3. Each person respected the input of the others. This was due partly to the personalities involved and partly to each having an important, individual expertise: marketing, sales, manufacturing, R&D, corporate planning, and accounting.

The team members were mostly in agreement that if they had to do it again, they would remain leaderless and schedule more frequent meetings with various outside consultants. The consultants could help them: (1) develop more creative ideas from within the group itself, (2) focus on the top priorities, and (3) resolve any current disputes.

For this company the venture team was an innovation in management practices: setting up a venture team with managers from around the company was in itself a departure from how they typically identified new business opportunities. The team's decision to be a multileader group was also an innovation in management practice. Although this group represents only one approach to venture teams—even Signode has tried many other approaches—it provides one good model.

Leadership and Delegation Styles

There is no set leadership style that is best for every situation. It depends on who you are, who you are leading, and under what circumstances. Although "participation" has been widely promoted—even in this book at times—it is not always desirable. Would you want a participatory style of leadership if you were a passenger in a nose-diving airplane? There are comparable times in the life of an organization when more directive styles are more appropriate.

In all, there are five major styles of leadership and influence:

1. *Tell:* "Based on my decision, here's what I want you to do."

2. *Sell:* "Based on my decision, here's what I want you to do because
. . . (benefit for listener)."

3. *Consult:* "Before I make a decision, I want your input."

4. *Participate:* "We need to make a decision together."

5. *Delegate:* "You make a decision."

Furthermore, there are three levels of delegation:

1. *Ask:* "Produce this result, and ask me before you take any action."

2. *Inform:* "Produce this result, and keep me informed of what action
you have taken."

3. *Do:* "Produce this result, and I don't need to know what you have
done."

In a creative climate for innovation there is an emphasis on *leadership* rather than *management/control*. At W. L. Gore and Associates new ideas are backed only when a team appears to support them. The philosophy of team development is that "leaders are established by followers."

As you recall from chapter 1, leaders empower people rather than motivate them. They provide environmental encouragement within which individuals can work creatively and productively. When "management" becomes "leadership," creativity can be managed without stifling it.

Leaders specify challenging outcomes that require individuals or groups to be creative in meeting their goals. They act more like team leaders, spending time developing vision and strategy, promoting interdepartmental collaboration, leveraging their people's talents, and building trust. They are willing to invest in and take risks with individuals. They go beyond passively accepting whatever comes through the suggestion box. They actively seek to use and benefit from the unused creative potential of their employees.

The new work ethic strongly values individual responsibility over following orders, personal expression of our unique talents (creativity) over "fitting in," and creativity over business as usual. When managers and employees representing different work ethics try to work together, their value differences can cause distressing problems.

One way of understanding value differences comes from SRI's Values and Lifestyles (VALS) program, which portrays three major groups of people and their predominant values:

1. *The Need-Drivens.* These are people so limited in resources (especially financial) that their lives are driven more by need than by choice. Their values center around survival, safety, and security.

2. *The Outer-Directeds.* These people conduct their lives in response to signals (real or imagined) from others. Consumption, activities, and attitudes are guided by what the outer-directeds believe others will think. Their values center around belonging and achievement. In general, outer-directeds are the most satisfied with the cultural mainstream of America, having created much of it themselves.

3. *Inner-Directeds.* These people conduct their lives more in accord with inner, private values rather than values oriented to externals. Their values center around self-expression, inner experience and growth, and social consciousness. They tend to come from outer-directed or need-driven families, having perhaps been satiated with external-oriented living.

A high-energy outer-directed person can be perceived by an inner-directed one as manipulative rather than politically astute, overly ambitious rather than goal achieving, and internally insecure/immature rather than as justifiably seeking rewards.

A high-energy inner-directed person can be perceived by an outer directed one as cocky rather than self-confident, ladder-climbing rather than self-expressive, disrespectful of authority (and needing controls) rather than visionary and persistent.

At certain times in my own life, my harmonies and disagreements with my own bosses have dramatically reflected such value differences. Such moments have been either the most touching or the most excruciating experiences in my work life. For all of us, I believe, learning to handle such value differences is as important as it might be difficult; in the past ten years, inner-directeds have grown from 4 percent to over 21 percent of the U.S. population, and the differences have become a major factor in organizational life. The differences need time to be voiced and to be seen open-mindedly and creatively. The values of outer- and inner-directeds can work well in co-SPIRITEDs—or they can spell misery for all.

No matter where you are in the hierarchy, you can promote the leadership style(s) that you believe are most effective in managing yourself and others around you. This is part of managing your environment as well as your boss.

Task-oriented Teamwork

Think about different work groups you have been in. To what degree did the group members do the following?

- Possess high mutual trust
- Set high standards
- Understand team objectives

- Talk freely
- Make use of member resources
- Make suggestions
- Relate their efforts to company goals
- Relate their group's work to that of other groups
- Work through conflicts rather than deny, suppress, avoid, or compromise them

Even if your groups practiced many of these qualities, you may have experienced some of the following problems that merit special attention:

- Splitting into opposing subgroups
- Lifeless group interaction
- Bowing to formal (or informal) rank
- Not sharing common goals
- Focusing only on the task or only on good relationships (imbalancing the group's effectiveness)

When a group is not working well together—not being as creative as they might be, for example—people often blame the interpersonal relationships: "I just can't get along very well with that person!" There are two general approaches to facilitating more effective group collaboration:

1. A relationship-oriented, open-ended, "let's see how we're relating to each other" approach, born out of the "T-group" (training group) experiences of the National Training Laboratories.

2. A more task-oriented, "let's improve our teamwork by focusing on a real project task" approach.

The T-group approach is most appropriate when "quality of relationship" is recognized by everyone as the key to improved teamwork. The task-oriented approach is most appropriate when team members are more focused on getting a job done.[1] A task-oriented approach also gives a repeatable process that you can train others in, a further benefit.

To develop task-oriented teamwork in both established and new work groups, you can explore five key teamwork issues:

1. *Mission/Purpose.* Does everyone agree on a single statement of the mission or purpose of the group?

2. *Roles.* Is there group consensus on what is expected of each person, especially in terms of responsibilities and authority?

3. *Procedures.* Are there mutually accepted ground rules for: resolving conflict, making decisions, communicating day to day, orienting new group members, and so forth?

4. *Interpersonal Relations.* After disagreements related to mission/purpose, roles, and procedures have been resolved, what else can members of the group do to live out a win-win ethic? Are there interpersonal "permissions" that need implementing to help the group stay on task and support harmony? Are there value differences that need to be accepted?

5. *Resources.* Are there sufficient human, time, and financial resources available to get the job done right?

These five constitute a hierarchy of issues. That means that the first issue should be explored and resolved with mutual agreements *completely* formulated before discussion is begun on the second issue. Likewise, the second should be completely handled before embarking on the third. This way is the safest for promoting trust in the group; the more general, broad-sweeping issues are handled first when trust may be the lowest.

The most personal issues—value, personality, and communication issues—are handled only after other issues have been resolved. Usually, the other issues are core to the disagreement, and the personal issues are few. In addition, by the time the interpersonal issues are discussed, there has been a chance to establish more trust during the previous discussions. As Irwin Federman, president of Monolithic Memories, puts it, "By trusting someone you invest in human dignity, in self-esteem, which cannot be purchased with money."

It is often wise to have a skilled, outside facilitator to act as a catalyst, guide, and arbitrator in resolving especially sticky issues. By taking one issue at a time and negotiating the best consensus agreement on purpose and goals, roles and authority expectations, procedures, interpersonal support, and available resources, any group can develop greater teamwork.

Alignment and Attunement

Alignment means that each person in your group affirms a stated purpose, vision, goal, or task. Attunement means that each person in your group shares in a communion, comradery, and feeling of personal support with every other member of the group.

There are many ways you and others in your group can help align yourselves and stay task-oriented. Some of your group will be better at some of these than others. The key is giving permission to yourself and others to do the following *actively:*

 a. Initiate topics and interactions

 b. Summarize discussions and decisions

 c. Clarify and elaborate on what people have said

 d. Take a straw vote to determine the consensus

 e. Exchange information and opinions

You and others can also help develop attunement in your groups to support its energy, goodwill, and unity. Again, some of your group will be better at some of these than others. The key is giving permission to yourself and others to do the following *actively:*

 a. Listen for understanding (not argument)

 b. Encourage others to participate

 c. Harmonize (without cutting short important differences of viewpoint)

 d. Ensure that everyone can participate safely (and be understood)

Alignment alone is not enough to respond to the economic realities of today and the future. With alignment alone we are caught living only through the designs of our intellects, disconnected from our hearts. Both alignment and attunement are important if our organizations are going to be more profitable and more responsive to serving all their stakeholders. The knowledge workers who are fast becoming a corporation's key assets are also becoming more and more insistent on working where attunement is strong.

Earlier we discussed love as an ultimate motivator for our creative spirits. Love can be translated into these two themes of alignment and attunement. Alignment gives us a single-mindedness of purpose. Attunement gives us single-heartedness of purpose.

Acting as a Parallel Organization

Sometimes—perhaps often—it may be difficult to influence changes in innovation practices through the normal lines of authority. The "system" may seem too strong for you to have any significant impact on it. This can occur especially when the primary goal of the organization is to do the same thing (producing products, etc.) in large quantities, whereas the goal of innovative thinking is to do some new thing, even if only once.

In those cases small groups have been effective in forming *parallel organizations* outside of the normal hierarchy. There are three purposes of parallel organizations:[2]

1. Accomplish new things by amplifying localized capabilities

2. Solve problems by task forces

3. Modify organizational culture and values through "transition management"

Effective parallel organizations have the following qualities and effects:

QUALITIES	EFFECTS
Broad representation (across functions and/or departments)	Breaking down isolation and opening up communication
Equal participation	Developing nonhierarchical relationships with flexible knowledge- and skill-based leadership
"Network" power	Creating new coalitions and a spirit of collaboration
New norms and values	Influencing by example rather than authority
Win-win problem solving	Diminishing turf battles

When you operate in parallel organizations, you bootleg resources from everywhere. You seek to find leverage points for exercising power through new alliances. Free of traditional barriers and constraints, you actively seek support from—and give support to—nontraditional sources anywhere in your organization. Parallel organizations are mentioned again in chapter 12 in the discussion of managing change processes.

IN CLOSING . . .

■ *Certain people I manage are computer programmers who are very smart and very technical. Some of them have very, very poor interpersonal skills. They can write a program but are very introverted.*

If you can get the programmer who has the good communication skills of a future manager in a partnership with the technical person, then you can send him or her to talk to a business person. The 'manager' will do most of the communicating, but he'll get his technical information from the technical person. If you're good at pairing them up, then the

technical person soon learns communication skills just by
watching, because he or she wants to be taken seriously in his or
her presentations.

Sue, project manager of computer systems in a bank ◤

Creativity and innovation do not happen in isolation. A SPIRITED individual may have particular strengths, yet from an organizational perspective it is equally important to emphasize collaborative roles. We must align and attune ourselves with each other if we are to have peak-performing groups (including large groups called organizations). Whether through informal roles, co-SPIRITEDs, leadership styles, or teambuilding, we must also combine our talents more closely. Only then can most groups bring an idea to final achievement.

Just as creativity in general is best nurtured by fluency in your logical and intuitive processes, so also are teamwork and innovation nurtured by the harmony of head and heart. "Feel with your minds, think with your hearts" is an appropriate motto for successful, innovative work.

CHAPTER 9

Taking the Lead in Group Problem Solving

"Everyday" Problem-solving Meetings

■ *We were talking in a meeting this afternoon about marketing—who was doing marketing, who was not terribly great at it. We kept asking, "How are we going to do better at marketing?" and it got more and more ponderous. We started asking different kinds of questions and brainstorming about how to market. "What did we do that was successful? What will work this year? How do we get the edge that we did in other great moments?"*

We found out that all the business we had gotten was from personal contact, not by calling them up. And all our personal contacts came from us doing free lectures and calling up the companies and saying, "We're doing a research project where we're trying to validate a questionnaire, do a workshop, collect data . . ." So we saw that marketing for our company should be to do the maximum amount of free events. We just switched our whole marketing from selling to giving. We figured if we each do fifty presentations this fall, we'll be rolling in business.

Dennis, clinical psychologist and consultant ■

How well does your day-to-day work group come up with alternative solutions to problems and evaluate them? Are you as effective as you might be in these group processes?

When imaginative new ideas either fail to materialize or get rejected inappropriately, it is often not because groups are necessarily inefficient, but because the group process was unnecessarily ineffective.

Perhaps the most immediate way to heighten the creative climate of our organizations is purposely to develop more creativity in our day-today work projects. These almost always involve meetings, whether "everyday" meetings or major problem-solving workshops (which might run three to five days).

Meetings take time, money, and human energy. Ten to 15 percent of most personnel budgets are for people's time in meetings. Middle managers typically spend one-third of their time in meetings. Top managers usually spend two-thirds of their time that way.

Improving meetings does not require suspending all operations and going to "organizational marriage counseling." By working to improve both the products and the process of our work simultaneously, the improvements we make will have a more immediate and enduring impact.

Most problem-solving meetings can use improvement, especially brainstorming sessions. Sometimes when your group gets together to work on some issue, you all might spin in circular discussions, never reaching a decision or even generating a productive set of alternative solutions. It's likely you leave these meetings with frustration on your faces and "What's the use?!" in your minds.

Do you recognize any of these common problems in your meetings?

- Unclear objectives and expectations

- Information overload

- Repetition and wheelspinning

- Straying from one subject to another without completion

- Talking at the same time

- Personal attacks

- Win-lose approaches in decision making

- Unresolved issues of roles, responsibilities, and authority

- Manipulation ("rubber stamp")

- Insufficient preparation and/or follow-up

Or perhaps your group has inadequate problem-solving methods. These are characterized by discussions that jump around in an unclear problem-solving process. This most often occurs if you:

a. try to solve the wrong problem,

b. state the problem so it is difficult to find creative solutions,

c. state the problem as a particular solution,

d. try to get agreement on a decision before having agreement on the problem.

How can you lead your group to find alternative solutions to problems and evaluate them? To improve problem-solving meetings, there are two major practices:

1. Agree on a specific *problem-solving process* and use it.

2. Establish clear *roles and expectations* (including decisions to be made).

Problem-solving Process

Problems are simply situations we want to be different. Problems are okay. In groups the method of problem solving needs to be made explicit and the problem defined in a way that is solvable.

Nothing tears down the will and creativity of a group as fast as unproductive "bitch sessions" disguised as problem-solving meetings. In my experience, if you go through each step of a problem-solving process one step at a time, you can get twice as much done, twice as effectively, in half the time (or close to it). One process to follow all the way through implementation is the P'DAGEDIE process. Be sure to follow each step in order:

1. *Perceive*	Describe your situation objectively. What's happening? Who's involved? What are they saying? (Also, determine who will be making the final decision: the chairperson? the group?)
2. *Define*	Determine a mutually agreed upon, neutral definition of the problem. (Getting agreement on the problem statement is 75 percent of finding the solution!) Make sure the problem statement is phrased in neutral terms that do not imply a specific solution. Determine criteria for the optimum solution.
3. *Analyze*	Describe the worst, best, and current states of the situation. List the forces that are acting in the situation to make it better or worse.
4. *Generate alternate solutions*	Gather all ideas without any criticism, judgment, or evaluation. "That's an idea . . . what's another?!"
5. *Evaluate the alternatives*	Discuss each alternative using the criteria from step 1. Discuss combinations too.
6. *Decide*	Choose your solution(s) using the authority of one person or of the whole group, as appropriate.
7. *Implement*	Create a plan designating who is responsible for what by when. Include checkup appointments and ensure that each person has authority commensurate with responsibility. Identify how you will recognize success of the plan when you see it. Then implement the plan.

8. *Evaluate*	Measure the success of your plan. If it didn't achieve your ex-
results	pected results, was it because of faulty implementation, inac-
	curate definition of the problem, or what?

To implement P'DAGEDIE properly your group should agree up front to complete one stage of the discussion before going on to the next stage. That way, everyone is working together instead of some defining the problem, others generating solutions, others analyzing the data, and still others debating the merits of a suggested solution.

Some of the questions you can ask to help in each of the P'DAGEDIE steps include the following:

Perceive:

How would each of us describe what is or what is not happening in this situation; who, what, where, when?

What are the best, worst, and most probable cases we can imagine?

How do we each feel about this situation?

How would we like it to be, and how is it different from that?

Who will be the final decision maker regarding this situation?

Define:

Is there a difference between the problem as given and the problem as understood?

What *isn't* the problem?

How can each word of our problem statement be made more specific?

Is this statement really a solution to a real problem?

Can the problem be diagramed?

To what extent is the problem within our domain of influence to resolve?

What are our criteria for an optimum solution?

Analyze:

What would a force field analysis show us?

What would experts say?

Can we break the problem into component parts?

What are all the facts?

Generate Alternatives:

What have others done?

What techniques can we employ to stimulate creativity (see chapters 5 and 6)?

Evaluate Alternatives:

How does each alternative or combination of alternatives rate by the criteria?

What rank or priority do we get?

Decide:

What would a straw vote show us?

What is the majority vote?

What executive decision is there?

Should we decisively delay making a decision?

Can we build up to a decision or eliminate alternatives to identify a decision?

Implement:

What milestones can best mark our progress?

Who will be responsible for what?

How can we gain the support of others?

Evaluate Results:

How will we know when we've been successful (or need midcourse correction)?

Do we need to redefine the problem and find new solutions?

How can we celebrate success?

You may recognize the similarity between this model and the APPEARE model discussed in chapter 4. Both methods start with being firmly grounded in present reality, build to a sense of vision, then generate ways of actualizing that vision. Both make use of any idea-generation methods described in this book. You may prefer one method over the other.

When the central purpose of the meeting is to generate creative alternatives, with the decision to come after evaluating the alternatives, I have found that success is more likely by following these guidelines:

a. Prepare in advance (creativity exercises, etc.).

b. Expect creativity from every area of your organization.

c. Involve people with diverse backgrounds (including some nonexperts).

d. Involve decision makers (to increase the chances of implementation).

e. Develop an agenda.

f. Establish the creative climate from the beginning.

g. Encourage contributions from everyone, but don't allow guests, observers, or onlookers.

h. Employ a variety of creativity techniques.

i. Move beyond "free-floating" brainstorming activities quickly (people tire easily of them).

j. Focus on the problem (rather than techniques).

k. Emphasize people's talents (rather than creativity techniques).

l. Avoid tying an idea to a particular person.

m. If a language barrier exists, allow subgroup participants to talk freely in their native tongue.

n. Encourage humor and "bizarre" statements whenever possible!

o. Spend twice as long generating ideas as screening them (typically).

Roles and Expectations

There is another major barrier to an effective idea-generation or decision-making meeting: the person with the most interest in the meeting's outcome is usually the one who directs the way the discussion flows. Most often, this is the manager.

When managers conduct their own meetings, certain difficulties typically arise:

a. They block participation by talking two to three times too much.

b. They find it hard to be nonjudgmental.

c. They focus on content and ignore process.

d. They find it hard to share decision-making power (or participants find it hard to take it when offered).

e. Staff end up with less buy-in, responsibility, and personal development.

Michael Doyle and David Straus wrote a wonderful book in the 1970s entitled *How to Make Meetings Work*[1] that first outlined the need for separate meeting roles for the following individuals:

a. The person with authority, in charge of calling the meeting—the *chairperson*

b. The person in charge of conducting the meeting—the *facilitator c.* The person keeping a written record of the discussions in full view of everyone—the *recorder*

d. The participants with vested interests in the discussion—the *group members*

Chairperson   The chairperson is responsible for the group performing its duties and is there to give the group direction in setting goals and plans, to provide expertise, to be the leader by authority, and to delegate. He or she is also responsible for participating in the discussions but not for guiding the process itself (see "facilitator" below).

Before a meeting it is important for the chairperson to specify what style of decision making he or she is intending for the problem at hand. Recall that in reaching and implementing decisions, the five major categories of leadership style are:

1. Tell: "Based on my decision, here's what I want you to do."

2. Sell: "Based on my decision, here's what I want you to do because . . . (benefit for listener)."

3. Consult: "Before I make a decision, I want your input."

4. Participate: "We need to make a decision together."

5. Delegate: "You make a decision (. . . and then ask, inform, or just carry out your decision)."

With the consulting style, the chairperson can also specify to what level in the P'DAGEDIE process he or she wants input: Only a definition of the problem? Analysis too? Alternative solutions? Evaluation of alternatives?

Trust in your group will be undermined if people believe the style is participative ("We decide") when the chairperson really intends it to be consulting ("I'll decide after your input") or selling ("Here's what I want to convince you to do").

Facilitator   A neutral facilitator only guides and monitors the process of participation by the chairperson and group members. Through the facilitator the climate is freed for more adventurous and productive interaction. The facilitator's role is much like that of an orchestra conductor: he or she doesn't play any of the music but keeps everyone working together effectively. In idea-generation sessions, he or she is usually most effective assuming a low-key yet catalytic role.

Being a facilitator takes quite a bit of skill in problem-solving processes and group interaction, but playing the role is enormously worthwhile. This is even more true the higher up you go in an organization.

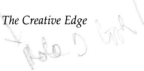

At higher levels the amount of time spent in meetings and the impact of decisions made there are higher.

Critical tasks of the facilitator are to:

a. identify how the meeting fits into the overall plan for dealing with an issue;

b. get agreement on a clear, achievable agenda;

c. ensure that the right participants attend (number, involvement, expertise);

d. set up a way for everyone to be properly prepared (especially the facilitator!);

e. plan the *process* for the meeting as carefully as the *content* to be discussed;

f. be specific about what decisions need to be made and summarize progress along the way;

g. clearly differentiate during the meeting between "Here's how we *will* discuss this topic" (process) and "Now let's begin discussing the topic" (content);

h. follow problem-solving processes (such as P'DAGEDIE) with discipline;

i. stay on one subject at a time (but make sure the off-the-subject comments are retained somewhere to ensure getting back to them);

j. use visual helpers such as a "group memory" (to be discussed);

k. use task- and support-oriented behaviors;

l. ensure that each person has a free and safe climate to give her/his ideas without being attacked but without taking over the floor.

Recorder The recorder provides a collective memory of a meeting by recording, on easel pads or newsprint, the main discussion and decision points. Through the recorder, the group memory:

a. focuses the group on problems and tasks (as displayed concretely on paper) rather than on personalities;

b. holds ideas to prevent information overload;

c. gives a psychological release point for participants, who then don't have to hold onto their own ideas;

d. prevents repetition and wheel spinning;

e. helps demonstrate and ensure that participants have been fully understood;   •

f. provides an instant record of the meeting's content, decisions, and delegated actions;

g. allows newcomers to the meeting to catch up quickly;

h. reduces status differential by giving equal weight to all participants' ideas.

Group Members   The group members are responsible for participating and facilitating as much as possible. They are the ones who have to implement and live with the outcomes. (Nonparticipation is also a choice, an exercise of personal power, albeit an abdication.)

When a meeting is over the most important and helpful way to improve meeting effectiveness is to take five to ten minutes and say, "Now that the meeting is over, how did it go?" Critique both the content and the process of the meeting. A simple way to do this is to discuss the following issues:

Listening.   Did people listen carefully to each other's views?

Openness.   Were ideas expressed with candor? Were differences thrashed out?

Task Orientation.   Did the discussion stay on track to get the task accomplished?

Participation.   Was there lively interplay, with many members contributing and no one person dominating the conversation?

Atmosphere.   Was the atmosphere satisfying, challenging, and stimulating?

Mutual Support.   Was there genuine concern for others?

Leadership.   Was the style of leadership appropriate? Was it concentrated or shared appropriately?

Task Accomplishment.   How well was the task accomplished?

Major Problem-solving Workshops

Sometimes, large-scale meetings are called to solve complex issues in productively structured yet creatively unstructured settings. For example, one Japanese electronics company wanted to decide how to invest their R & D budget to be properly positioned for opportunities that might arise in the year 2000 and beyond. Their interests ranged from electronics to new materials to biotechnology. At another time, a consumer health-products company wanted to see how some of the more sophisticated technologies, such as ultrasound, might be used in producing their products.

Other examples include:

- Helping one division of a chemical firm establish a broader vision of the nature of the business (including specific business opportunities)

- Developing a financial institution's business strategy for the entire Asia-Pacific region

- Identifying business opportunities for 1990–95 with a Japanese company in the software industry

- Formulating new business opportunities for using excess engineering personnel in a nuclear engineering firm

- Inventing a new portable printer

These workshops often bring together a diverse group of specialists and require workshop management as complex as the topics themselves. A look at how these are conducted will give you a feel for how to apply effective meeting ideas to these workshops.

There are four stages, with eight steps, to producing such workshops, often called Innovation Searches (or Business Opportunity Searches).[2] These eight steps are shown below.

There are eight steps to producing an Innovation Search (or Business Opportunity Search):

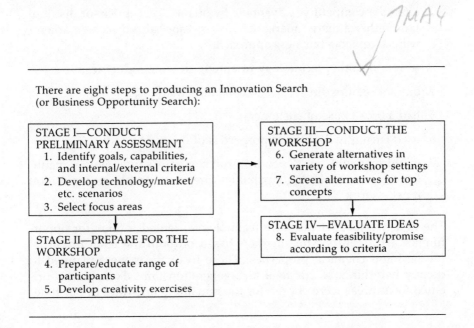

STAGE I—CONDUCT PRELIMINARY ASSESSMENT 1. Identify goals, capabilities, and internal/external criteria 2. Develop technology/market/ etc. scenarios 3. Select focus areas	STAGE III—CONDUCT THE WORKSHOP 6. Generate alternatives in variety of workshop settings 7. Screen alternatives for top concepts
STAGE II—PREPARE FOR THE WORKSHOP 4. Prepare/educate range of participants 5. Develop creativity exercises	STAGE IV—EVALUATE IDEAS 8. Evaluate feasibility/promise according to criteria

Stage I—Conduct Preliminary Assessment

First you develop the groundwork for the *content* of the workshop by meeting with the principal stakeholders of the project. Develop a consensus regarding the following:

Objectives. What are the objectives of the overall project? What are the objectives of the Innovation Search workshop in relation to the project objectives (how does it fit into meeting the project milestones)?

Criteria. What are the general parameters by which to focus the Search?

Capabilities. What are the organizational capabilities and resources—financial, technological, human, and so forth—upon which to build the ultimate solution?

Background Information. What information is needed to provide a substantive groundwork for idea generation—market and technology trends, for example?

Focus Areas. What are the topics by which to organize and guide the idea-generation discussions?

For example, a chemical company wanted to identify new applications for a resin product they had invented. As criteria to guide their search, they wanted business opportunities that:

1. could be commercially developed for either the short term (three to five years) or long term (five to ten years);

2. offered a competitive advantage by quality, cost, price, or distribution (either a "large market share" or "specialized niche markets" were acceptable business approaches);

3. were in healthy, technology-oriented, growing industries;

4. did not require direct sales to consumer markets;

5. had a broad base of customers;

6. used currently developed properties of the resin product;

7. might or might not depend on the company's engineering manufacturing, or marketing capabilities (they would make an acquisition if necessary).

The first two focus areas were technology-oriented to look for opportunities across all possible markets based on both the resin products' physical and chemical properties. These focus areas were planned to occupy two-thirds of our total idea-generation time. Four market-oriented focus areas were chosen for the remaining one-third of our time.

This ensured that we would consider the needs of four key industries that might use the product.

As another example, the project with the Japanese electronics company began by constructing four alternate scenarios for the year 2010. We first examined each for the range of market needs that might be expected. Those market needs were combined in a matrix with technology trends, including: biotechnology, advanced materials and processing methods, fuel cells, optics, electronics, electro-active and "smart" polymers, specialty chemicals, membranes, robotics, artificial intelligence, digital communications, optical electronics, chemical synthesis, energy conversion, and software.

From this analysis of market possibilities and emerging technologies, plus a knowledge of their corporate capabilities, focus areas were selected to organize the Search workshop.

Stage II—Prepare for the Workshop

Next you prepare the groundwork for the *process* of the workshop:

1. *Select participants.* Who are the best people to attend based on (a) specialist expertise, (b) importance for eventual implementation, and (c) comfort/skills in group settings? (These may be people from outside your group and even outside your organization.)

    In the resin product search the participants included the company's project manager, new venture manager, director of research, research and development manager, market development manager, the product inventor, and another senior scientist. We supplemented this talent with specialists in particle technology, catalysts, membranes, and the four target industries.

2. *Prepare participants.* What articles, technology briefs, market analyses, or other materials would help the participants to arrive at the workshop on a common ground, ready to work closely in multidisciplinary problem solving with people of different specialties? If presentations will be given (on market trends, etc.), what preparation and coaching do the presenters need?

    For the resin-product Search a workbook with over 400 pages of articles on various industry trends and technology developments was prepared for all participants. Although each person didn't read every page, it helped give everyone a common vocabulary for the workshop. Video and other media are also excellent tools for such preparation.

    The presenters were coached to give a twenty-minute talk on pertinent subjects designed to *stimulate* the group, not to solve the problem. Roughly five minutes were devoted to an orientation to a subject, ten minutes to details important to our idea generation, and five minutes of specific initial ideas.

3. *Prepare the search facilitator/manager.* What depths of knowledge does the facilitator/manager need to absorb to manage expertly the participants' preparation and workshop interaction?

This preparation always includes reading the participants' preparation materials. It may also involve activities such as interviewing each participant in person before the workshop.

4. *Prepare the search agenda and idea-stimulation exercises.* What is the optimum structure of focus area discussions (and presenters if appropriate) for the agenda? What linear and intuitive techniques might best stimulate ideas in each of the focus areas? What combinations would keep the flow going? Which ones would work best with the whole group, with subgroups, and/or individually?

The resin products search included exercises such as matrix analysis, analogy, reframing questions, force field, and brainstorming.

5. *Prepare graphics and "show and tell".* How can the Search workshop environment be dressed up to help focus the imagining and enliven the climate for creativity?

Stages I and II can span a period of four to eight weeks to give time for each person to prepare adequately.

Stage III—Conduct the Workshop

Managing the Search workshop usually requires a skilled facilitator (or facilitators) and a second (or third) person as recorder and subgroup leader. This allows the key stakeholders to participate freely in the discussions while the group process is being properly guided.

By following the guidelines covered earlier in this chapter, the facilitator and recorder may become rather invisible to the workshop. The attention should remain on the quality and quantity of ideas being generated. This is as it should be, though an observer might misconstrue the facilitator's limited participation.

After introductions, start the Search with a discussion of objectives and criteria to make sure that participants have a common understanding. This also gives the key stakeholders a chance to answer questions and modify or amplify their needs. Then be sure to set the climate for creative interaction by introducing ground rules such as the following:

- Give all your ideas—even wild ones.

- Give your ideas freely and spontaneously (the facilitator can guide this).

- Don't wait for the "right time."

- Share your ideas with everyone—avoid side conversations.

- Draw or sketch your ideas if you desire.

- Make sure your ideas and others' are recorded properly.

- Build on others' ideas—avoid idea killers.

- Enjoy, have fun, and use humor!

Then proceed with your agenda, including the creativity exercises, and employ other techniques as the need arises. As ideas are recorded, periodically have them put into a word processor or computer so that a printout will be available for reviewing the ideas. Make sure each idea is assigned its own number, no matter what focus area it was recorded in.

In a two- or three-day Innovation Search workshop, you might record 300 to 600 separate-and-overlapping ideas for all the focus areas. There are many ways to group the ideas into concepts (a "concept" might represent anywhere from one to twenty-five or more "ideas") and do a preliminary screening of the criteria. One process involves the following five steps:

1. *Review the criteria for the Search.*

2. *Have each participant skim all the ideas,* briefly marking the ones that strike his or her fancy.

3. *Have each participant go through his or her marked ideas and begin assembling them into concepts* that represent possible answers to the workshop problem/objective. Document each concept with (a) a title, (b) a description, and (c) reference numbers to individual ideas included in the concept.

4. *As a full group, have each person presenting one of his/her concepts.* Ask others if they have a similar concept they would like to add/merge with the presenter's. Give the presenter veto power over whether he or she wants the others' concepts added (this stops any tendency to argue rather than build on concepts). Any concepts that were offered by others and vetoed by the presenter may be introduced at another time. Move on to the next person and continue this process until all concepts have been presented. My experience has shown that, on the average, if 10 people each begin with 10 concepts (a total of 100 concepts), approximately 40 concepts will emerge from the synthesizing process.

5. *Apply some ranking method to specify which concepts, in the wisdom of the group, most deserve a more detailed evaluation to determine their attrac-*

tiveness in meeting the objectives. In some circumstances, this ranking can provide a final decision, but usually a much deeper level of investigation is needed. One quick method for ranking is to have each participant assign two points to each of her/his "top third" of all concepts, zero points to each of her/his "bottom third," and one point to each of her/his "middle third"—based on the criteria. Then add everyone's point assignments to produce a prioritization.

Stage IV, evaluating the ideas more fully, is discussed in detail in chapter 11.

IN CLOSING . . .

> ▟ *The car frame has 4,000 spot welds, which distort the frame in some way. Everyone knows this, but no one admitted it. You can distort the steel parts that must fit on the frame, and it looks reasonably good. But plastic parts are so accurate, they need a perfect body to sit on. Our design had separated the plastic body and the steel frame, but we couldn't put the two together.*
>
> *We had one of those three-day-and-night, staying-in-the-office meetings and suddenly found the solution. All we had to do, first, was admit that there's no way on God's earth to put 4,000 welds on and not distort something. When we admitted it, 90 percent of our problem was solved. The solution was very simple. We had to do exactly what the casting industry does. You make the casting larger than the part: you put the part there and machine it.*
>
> *Hulki, chief design engineer for automobiles* ◀

Meetings have an incredible impact, not only on *what* gets accomplished in your organization, but also on *how* it happens. Productivity, morale, teamwork, and commitment all are affected. Usually it takes much longer to recuperate from a poor meeting than it takes to hold it in the first place.

Effective meetings, especially for creative idea generation and problem solving, can be a real pain or a real pleasure, depending on the skill and sensitivity with which they are conducted. Managing your meetings properly can make a huge difference in the emotional energy you invest in seeing an idea through to completion. Conducting your creative sessions with solid preparation and a variety of idea-stimulation techniques can enhance the quality and quantity of alternatives arising from the very different perspectives in your group.

It takes practice and diligence, but it's well worth it. Go for it.

CHAPTER 10

Business Innovation with a Purpose, Vision, and Strategy

What Do We Stand For?—To Be or Not to Be

◖ *I had a boss who was a real visionary. She used to get up in meetings and talk about her vision for the nurse practitioner program, about her image of health care.*

We replaced the medical model with the self-care model. We replaced it by going inside ourselves and understanding our own self-care process. We would work weekends and nights and drive to different cities, but we were all really alive.

We went through a staffing process that was transformational for staffing in a medical school, producing a values statement about who we were and what we were doing. It gave us an intense sense of belonging and commitment. People who didn't like those kinds of values and vision left, which was fine. It wasn't a wrenching leaving; they just got very clear that this was the wrong place for them.

We were invincible. I think part of learning about your own creativity is pushing through what you think are your limits. I think that's what creativity is all about.

Cynthia, consultant and former health-care professional ◗

Where is your life going? What do you stand for? When do you feel most purposeful, productive, and attuned at work and generally in life?

These questions were first asked in chapter 4 in relation to "Persistent Vision" in the APPEARE process. Without a sense of *purpose* and *vision* about our lives—a sense of self—we may feel adrift, constantly reacting to changes in life without much sense of power. The same is true of our work groups and our organizations. This chapter can help you enhance your CREATIVE climate for innovation by focusing on "Vision and the Role of Innovation" and "Environmental Monitoring."

How would you and those you work with answer the following questions?

Q What is the essential business of our organization?

Where are we going?

What do we stand for?

When are we most alive, purposeful, productive, and attuned to our business environment?

Your organization's purpose answers the questions, What do we stand for? What are our values? In the long term, what do we want to be? What are we *called* to be? What is our business and our ultimate contribution to the society and economy we work in? For example, John Sculley, president of Apple Computer, has said, "Our dream is to change the world. Now that's a pretty bold dream, but if you are going to have dreams, you might as well have bold ones."

As a result of a "strategic visioning" conference with an electronics firm, they drafted a statement that read, "Our purpose is to truly serve the information management needs of targeted customer groups with the highest levels of technical/market leadership, reliability, customization, and collaboration. Continuing high growth and profits are significant measures of how well we are achieving this purpose. Internally, we operate by, and expect of each other, equally strong commitment to these same values of leadership, reliability (trustworthiness and honesty), customization (flexibility and personalization), and collaboration (team spirit)."

Your organization's *vision* is a description of the medium-term (five years, typically) implementation of its *purpose*. Ask yourself and your organization, "What will we be producing in five years? How will we be organized? What values will we be emphasizing? What is the role of innovation for us (in products, marketing, manufacturing, etc.)?"

The vision may remain the same while particular financial, product, and human resources goals may change. Or the vision itself may change every few years. Throughout such changes, the purpose provides a fundamental continuity and stability for your organization.

Thomas Watson, Jr., son of IBM's founder, described the importance he placed on shared purpose and vision this way:[1]

> I believe the real difference between success and failure in
> a corporation (is) how well the organization brings out the
> great energies and talents of its people. What does it do to

> *help these people find* common cause *and* sense of direction *through the many changes which take place from one generation to another? . . . If an organization is to meet the challenge of a changing world, it must be prepared to change everything about itself except (its core) beliefs.*

In *In Search of Excellence,* Peters and Waterman spoke of the principle of "Hands-On, Value-Driven . . . in touch with the firm's essential business."[2]

> *Let us suppose that we were asked for one all-purpose bit of advice for management, one truth that we were able to distill from the excellent companies research. We might be tempted to reply, 'Figure out your value system. Decide what your company stands for. What does your enterprise do that gives everyone the most pride? Put yourself out ten or twenty years in the future: what would you look back on with greatest satisfaction?'*

Even work groups can have purposes. For example, Phil Turner, manager of Raychem's U.S. Facilities Department, once said, "We are a department that uplifts people's spirits."

Ultimately, a statement of purpose defines success for your organization. Dr. W. A. Pieczonka, president of Linear Technology, points out that success means different things to different stakeholders:

- For the *corporation* success can include wealth generation, contribution to society, leadership, and recognition.

- For the *founders* success can include affluence/security/independence, contribution to society, fulfillment, the thrill of winning, and work as leisure.

- For *employees* success can include a piece of the action, challenge, recognition, fulfillment, and work as leisure.

- For *customers* of the organization success can include security, value performance, and a competitive edge.

- For *society* success can include jobs, a clean environment, needed goods and services, and a strong economy.

Signode Industries, Inc., recently spent months gaining consensus among its executives on a new statement of corporate purpose. The process has produced more than just words on paper. The executive team is now working more closely than ever to take their company into the next decade as "the world leader in a broad range of packaging, fastening, securing, and control/identification applications," with unfailing commitment to understanding the marketplace and to providing products and services of unequaled value. By this course they aim to be recognized in the top ranks of the world's best-run companies. High energy, hard work, integrity, and dedication are highly valued at Signode; and beyond this, certain factors are seen as critical to their success: a humanistic working environment that allows their people to feel fulfilled, a strong entrepreneurial spirit, and cooperation, support, and networking.

Notice the clarity in one chemical company's statement of objectives and philosophy:

> It is the primary objective of ———— to produce maximum long-term profit . . . (to) benefit the company's stockholders, (to) benefit the company employees by enabling us to offer them greater opportunities for self-advancement and improved job security, (to) benefit the company's customers by enabling us to provide them with greater services and better products, (to) benefit the general public and the economy at large by enabling us to contribute more broadly and continuously to the improvement of living standards throughout the world.
>
> We will work toward several corollary objectives necessary to the achievement of our primary objective:
>
> We will find, attract, and hire the most imaginative and competent people available, and we will treat them well in every respect, specifically including compensation commensurate with their contributions.
>
> We will provide the equipment and training necessary to achieve maximum productivity with maximum safety.
>
> We will market our products and services based on the optimum in price, quality, service, and delivery; and we will be honest in our claims and advertising.
>
> We affirm that continuous innovation (author's emphasis) is the primary basis of our profit growth, whether it be accomplished through process improvement, through new products or new uses for existing products, through new ways of marketing products, and through basic research.

How well our organizations live up to their pronouncements varies widely, depending on the values and ethics of individuals. However, without statements of purpose and vision (a sense of self), there is nothing to measure day-to-day decisions by; constructive discussions can hardly begin. Without purpose and vision, our organizations remain reactive and off balance with their competition, pursuing short-term (one to three years) goals without a unifying direction. You can sense their floundering as they are tossed about by shifting market, technology, and economic forces in a rising storm of ever more rapid change.

Innovation remains important in *all* areas of your organization's life. However, the emphasis for innovation may shift depending on the growth cycle of your industry and products. For a while, when costs are the driving force behind competitive advantage, innovation may be aimed inwards at production and support efficiencies. Another time, innovation in new products carries the torch and gets the resources. At other times marketing innovations are the keys to success. Only with a purpose and vision in hand (and heart) can goals and objectives begin to make sense to people throughout your organization and the role of innovation remain clear.

Growing with Your Purpose and Vision

Even if your senior executives have a consensus on purpose and vision, and even if people in your organization understand what it is, there is still the question of *sharing* the purpose and vision. If you are like most people, you really want to produce quality work for an organization whose purpose is meaningful for you. When we feel a connection between our own values and purpose, the organization's values and purpose, and how we can express them through our jobs, our motivation skyrockets.

In one case I consulted with the board and staff of a shelter for battered women. Half of them believed their organization was established solely to provide a shelter for women in need. The other half believed they were also to provide community education. Although these two notions may seem very close, conflicts arose when resources were short—for example, when there was a staffing shortage at the shelter and someone was committed to speak to a community group. Only when the second purpose was agreed to, and only when a clear set of priorities for using time, people, and financial resources was established, was the group able to align itself to deliver its services and to feel attuned as an internal community once again.

How does a *shared* sense of purpose and vision emerge? In part, it takes a dedicated executive who is capable of *visionary leadership*, a "corporate hero." As stated by Deal and Kennedy in *Corporate Cultures*,[3]

> *We are not talking about good "scientific" managers here.*
> *Managers run institutions; heroes create them.*
>
> *The one quality that more often than anything else marks*
> *a manager is decisiveness, but heroes are often not decisive;*
> *they're intuitive; they have a vision.*
>
> *Heroes . . . are driven by an* ethic of creation. *They*
> *inspire employees by distributing a sense of responsibility*
> *throughout the organization.*

There is more tolerance for risk taking, thus greater in-novation; more acceptance of the value of the long-term pro-cess, thus greater persistence; more personal responsibility for how the company performs—thus a work force that iden-tifies personal achievement with the success of its firm.

The success of these visionaries (like Henry Ford, John D. Rockefeller, William Kellog, Harley Proctor and others) lies not only in having built an organization but also in having established an institution that survived them and added their personal sense of values to the world.

Successfully instilling values into an organization tends to come from sincere and sustained devotion to long-term values (of founders or other heroes) rather than from charismatic leadership. We can make ourselves into effective leaders. Personal commitment rather than personal mag-netism is the key.

A strong, visionary leader is vital to your organization's members working together to fulfill a purpose and vision. Yet the presence of such a leader is insufficient: the purpose and vision must be *shared*. Your top managers must extend themselves and communicate with employ-ees to find out what *employees* believe and want the organization to stand for. To ask that question virtually always elicits a statement of high val-ues, for which employees would give 100 percent if the organization's purpose were truly defined and practiced that way.

Such an approach is in the mode of a "consultative" leadership style ("I need your input before finalizing"). This is different from styles of "telling" ("Here is *the* vision for which you will work"), "selling" ("Here is a vision . . . let me convince you that you should want to work for it"), or "participating" ("*We* need to arrive at a decision about vision jointly"). Although the other styles are sometimes necessary, the con-sulting style usually has the advantage in balancing time, commitment, and business needs most easily.

It would be a mistake to think that everyone should sit around wait-ing for the CEO to meet with them and set the visioning process in motion. If you or anyone feels strongly about what you want for your organization, you can:

a. describe your own life's purpose and vision;

b. develop your notion of your organization's purpose and vision;

c. involve others to develop a shared purpose and vision;

d. disseminate and build the vision—discuss it, get feedback, and en-roll people in the highest purpose and vision you're willing to com-mit to;

e. design structures to communicate and implement the vision—stra-tegic objectives, accountabilities, ground rules, and systems;

f. produce results and guide your organization's evolution—tell the truth about where you are and where you're going.

In other words, enjoy and believe in what you do, and communicate that. Enroll others in your vision and enroll in theirs. You might even use images and metaphors to share your vision more powerfully.

Beyond leadership and initiative, a shared sense of purpose and vision also requires asking the right questions. Sometimes when people in an organization begin to discuss their vision, they start with the questions, "What are the business opportunities?" and "What does the market want us to be?" These questions play a critical role in business success. However, basing your organization's purpose and vision primarily on external forces keeps your organization in a reactive/responsive mode with no core stability and identity. These questions by themselves provide impractical answers for building a vibrant and profitable business.

Purpose and vision are born from a recurring discussion of five questions:

1. What do we want to be?

2. What does the marketplace want us to be?

3. What can anyone be?

4. What do our capabilities lead us to be?

5. What are we called to be?

This last question is aimed at the higher values of some executives that dramatically shape strategy decisions. For example, a midwestern insurance company once had a vice-president of marketing who advised the president repeatedly that certain marginally profitable groups should be given dramatically higher rates or be cut. The president, however, believed that his company's mission was to stand by the "family" of corporate policyholders. The family won out, and the marketing V.P. is now working elsewhere.

The practicality of the purpose and vision comes from a balance in answering these questions. For example, to redefine "What we want to be," a division of a chemical company first wanted to explore what the market might want. We worked with them by first exploring future scenarios of technologies and consumer needs. After generating ideas and screening them in a three-day Innovation Search (see chapter 9), they were asked, "If you were put in charge of any single business concept discussed in these three days, what would you want to invest your work energy in?" This initiated the discussion of "What do you *want* to be?"

Based on their answers and the three days of discussion, they could begin to see into themselves and their environment more clearly. With a subsequent detailed evaluation of key business concepts, they broad-ened the nature of their business from "spun-bonded fibers" to "non-woven engineered sheet structures."

A shared purpose and vision have become, more than ever, the pri-mary means of coordination and control over an organization's output. Normal goal setting, financial targets, and controls don't do the trick. Business plans that set major goals such as "position the net interest margin" rarely give nonexecutives clear guidance for their day-to-day decisions.

Decentralized decision making requires an emphasis on coordina-tion rather than control. With a shared purpose and vision, senior man-agement can be more assured that local decisions with customers, ven-dors, R & D personnel, and others are appropriate to organizational strategies.

What's Going on Out There?!—Balancing Purpose with Market Demands

"What do we want to be?" "What does the world want us to be?" "What are we called to be?" If any of these questions is not seriously addressed, your organization can seriously lose its balance. To only ask, "What do we want to be?" often results in a cocoon of product development typ-ical of the technology-driven firm that develops something in R&D and then asks marketing to find a customer. To only ask, "What does the world want us to be" often results in a scatterbrained reactivity to mar-ket trends. To only ask, "What are we called to be" results in organiza-tions with staff who are burned out from impractical, unproductive zeal.

Therefore, to set appropriate strategic direction, we all need to com-bine purpose and vision with understanding the external environment: the gusts and waves of changing social, economic, competitor, techno-logical, environmental, political, and human conditions. O. D. Re-sources, an Atlanta-based consulting firm, states its vision after clearly outlining the *environmental context* for its vision:

> *Human existence has reached a point in its evolution where the matrix of forces affecting day-to-day and long-range activities has become so complex and turbulent that learning to manage change is now a critical skill for sur-vival, not to mention, an important aspect of growth and prosperity. Knowing* what *(decisions to make) is no longer enough. Without the strategies for coping with constantly changing and increasingly sophisticated environments, the quality of life . . . is in jeopardy.*
>
> *Our vision is to be positioned as the dominant resource providing change-related implementation methodologies,*

training programs, and consulting support to decision mak-
ers throughout the world who are attempting to design new
paradigms for the betterment of human existence. It is our
belief that only with this level of "dominance" and "world-
wide" visibility will we be able to significantly influence the
human capacity to accommodate large-scale change.

In a recent survey of the major concerns of 188 CEOs from forty-four industry categories, the strategic issues they most often named were the following:

a. Fierce competition—domestic, international, and cross-industry

b. Deregulation of industries—competition, changing markets, and industry restructuring

c. Changing markets—market segmentation, consumer values, and offshore markets

d. Economic/financial uncertainty—growth/stability, inflation/interest rates, capital availability, and the international monetary system

e. Government policies/uncertainty—taxation, regulation, trade, and political risk

f. Technological impacts on business—tracking and timing of investments

g. Changing workforce—productivity, education/training, and succession planning

Gathering and disseminating information about these issues, in usable form, provide the greatest stimulus for innovation at all levels. This frequent stimulus keeps us thinking, feeling fresh and on our creative toes. Since it is often the case that the nonexpert on a subject can most easily see the nonobvious question or insight, it makes sense to distribute condensed monitoring information widely throughout the organization and provide a means for new ideas to surface.

Toshiba in Japan, for example, employs over a dozen full-time junior managers to monitor specific aspects of their business environment worldwide. These people then analyze and condense the information and distribute it throughout many levels of the organization. Even supervisors and technical specialists have direct access to the output of this monitoring system. They are stimulated by it to think of new, innovative ways of doing business. At a minimum, they understand more fully the context behind many executive decisions that might otherwise appear arbitrary, such as the introduction of cost-cutting measures or the funding of one project but not another.

Many managers may not see the need for widespread distribution of environmental information. Yet companies that do, such as Toshiba, broaden their base for sources of new ideas. In fast-moving markets, this broad participation in innovation can be the primary competitive advantage of your organization.

On the whole, monitoring the environment and distributing the information widely can:

- provide early notice of threats or opportunities;

- increase personnel's awareness of the changing environment and the implications for their own work;

- increase interdepartmental cooperation as well as bottom-up communication;

- assist early identification of promising, innovative individuals.

There are two approaches to gathering information about the external environment: scanning and monitoring. Scanning is an early warning system used to detect signals of new trends of potential importance to the organization. Monitoring is the detailed tracking of events and trends of known importance to the organization.

A system for scanning and monitoring not only ensures that an organization has up-to-date information about trends of known importance, but also keeps it from being surprised by new trends. By using this information to develop alternative scenarios about the future (as discussed in chapter 5), organizations can develop appropriate business strategies and hone internal operations.

Scanning Systems

A scanning system analyzes disparate events in the environment hoping to identify an emerging trend of significance for your organization. Such a system is not a luxury but a necessity. It provides you with security from surprise threats as well as insights into potential competitive advantages.

There are some good models for formal scanning systems. For example, the U.S. Congress, Bell Canada, and SRI International have all used a Trend Evaluation and Monitoring (TEAM) approach to identify emerging issues and early signs of social, political, economic, and technological change that could result in threats or opportunities.

TEAM is based on three levels of participation:

1. A volunteer group of monitors regularly reviews one publication or more (from a select list of appropriate literature) for items suggesting a change that could affect the organization.

2. Brief abstracts on those items are forwarded to a small committee chosen for their diverse backgrounds and analytical skills. Once a month they attempt to piece together various bits of information that could be important to the organization.

3. The results of the analysis committee's discussions are sent to a steering committee of senior management and to each monitor. The senior committee decides which topics to bring to the attention of specific managers or the organization as a whole.

The TEAM approach can be visualized in this way.

"TEAM" PROCESS

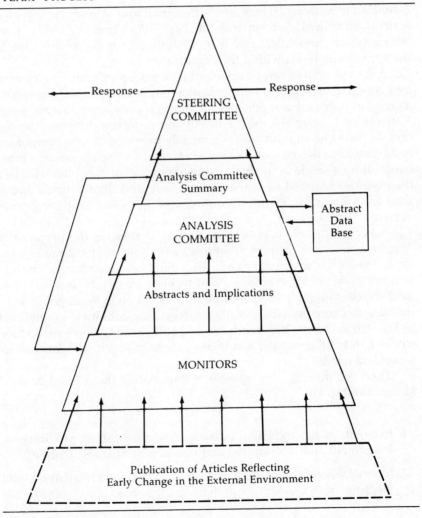

Monitoring Systems

Monitoring alone covers only the known driving forces affecting an organization's future.

There are four central issues regarding monitoring the external environment:

1. Establishing formal and informal systems

2. Targeting the information you want

3. Analyzing and interpreting the information

4. Applying the information in decisions and action

Establishing Systems   There are both formal and informal monitoring systems. In many U.S. companies, it is difficult to find any formal monitoring system, much less one that carefully distributes information to the appropriate levels within the organization.

A formal system can be operated by a single individual, a centralized component within the organization (such as marketing research), many decentralized components, or through coordinated networking. Networking is often the most efficient and economical monitoring option. In this case, coordination is generally provided by one component in the organization. Beware: the scope of a monitoring operation is often too great for a single individual; a single component may often also lack the expertise required to properly manage it; and decentralized operations are sometimes difficult to control, and they often produce gaps or overlaps in information.

A key feature of successful monitoring systems is the presence of "gatekeepers," (see chapter 8) who act as monitors of change outside the framework of a formal system. They might communicate with their technical cohorts, for example, about technology trends. Internal technical societies might hold gatekeeping colloquia on projects going on in the various company labs. Another gatekeeping activity is exemplified by Hewlett Packard where hardware and software engineers spend time working in retail computer stores to be more in touch with customers' wants and needs.

There are three general sources of information that should be used for monitoring:

1. External media, including public press, competitors' publications, investment analysts reports, and research companies' analyses

2. Internal reports, including technology and market analyses and feedback from vendors, customers, trade shows

3. Networking, including exchanging ideas within formal and informal groups (inside the organization and within professional groups)

To be worth the effort of monitoring, continuous reporting is necessary. "Continuous" means that the system produces regular reports of information, *immediately* points out deviations from key strategy assumptions, and alerts people to new and important trends. Attempts at "exceptions only" reporting erodes the usefulness of the information for updating the ongoing strategic and operational planning efforts.

Depending on the scope and complexity of the monitoring (and scanning) activity, an organization may want to computerize the information.

The combination of a formal system, gatekeepers, and other monitoring methods gives us maximum exposure to our external environment. We can experience a significant loss of valuable information if any method is missing.

Targeting the Information   There can sometimes be so much information, leading to so much confusion, that the monitoring may not seem worth the effort to us. Early on it is important to identify the type of information that is most relevant to our organization's key factors for success.

The information also needs to be formatted and reduced into a digestible form. When using the scenario method discussed in chapter 5, key driving forces can be identified relative to a decision. Information about these key driving forces can become the focal points for monitoring activities.

For example, suppose that we were concerned about who would most likely buy some particular highly innovative new product. Defining our consumer market and knowing how market segments respond to innovation would be very important to us. Our *initial choice of market segmentation schemes* could make or break our efforts to get relevant information.

One segmentation scheme with a good track record for introducing innovations to the market has been the Values and Lifestyles (VALS) typology developed at SRI, which was described in chapter 8. The VALS typology has been hailed as one of the top ten breakthroughs in understanding consumer behavior in the last forty years.[4] The VALS program staff have identified the values-types of those who are the keys to early and later adoption of new innovations. VALS identifies three major lifestyle orientations in the U.S. population:

1. The Need-Drivens, whose values center around survival, safety, and security.

2. The *Outer-Directeds,* whose lives revolve around achievement, consumption, and fitting in with others.

3. *The Inner-Directeds,* whose primary concerns are inner experiences and personal growth.

Within the three large groups there are a total of nine subgroups. One of the outer-directed subgroups is known as the Achievers (21 percent of the U.S. population), who include many leaders in business, government, and the professions. They are competent, self-sufficient, materialistic, hardworking, oriented to fame and success, and lovers of comfort. They are the affluent who live the American dream, so they are defenders of the economic status quo.

The three inner-directed subgroups are known as the I-Am-Me's, Experientials, and Societally Conscious. I-Am-Me's (3 percent) are in a short-lived transition from outer to inner directed. They tend to be young, very independent, perhaps even narcissistic. They often define themselves by their actions rather than their words. Experientials (6 percent) are probably the most passionately involved with other people, as well in as many social/human issues. They tend to go after direct experience and involvement in life, being very experimental and highly participative. The Societally Conscious (12 percent) have extended their inner-directedness to the society as a whole, even to the planet. Their sense of social responsibility leads them to support causes such as environmentalism and consumerism. They tend to be activists who are highly knowledgeable about the world. They also tend to live lifestyles of voluntary simplicity.

For the adoption and diffusion of new innovations in society, Everett Rogers at Stanford University has identified five innovation types: Innovators (2.5 percent), Early Adopters (13.5 percent), Early Majority (34 percent), Late Majority (34 percent), and Laggards (16 percent). Depending on the industry and the product, different VALS groups take the lead in actively accepting untried, highly innovative products and services. Most typically, the Innovators are more likely to be I-Am-Me's and Experientials, while Early Adopters are most likely to be Achievers and Societally Conscious.

Knowing this type of information in detail would help you to manage the marketing of your product and service innovations. One example of a company acting on targeted monitoring information is a home-development firm in Texas. Using the VALS system to analyze the home real estate market, they identified a certain type of Achiever and designed homes and advertising specific to those Achievers' tastes. Even during a downturn in the home buying market, their home designs were selling at the rate of 1,000 per year, propelling them into the fastest growing home-development firm in the state.

Analyzing and Interpreting Information   The wealth of available information is useless unless it is analyzed and interpreted. One application of trend information that can easily stimulate innovation is Vulnerability Analysis, where potential threats are examined according to the grid found below.

This is a powerful tool for highlighting the need and possible directions for innovation. The basic steps of Vulnerability Analysis are as follows:

1. Identify underpinnings critical to your organization's health—resources/assets, relative costs, customer base, technologies, competition, social values, sanctions, and so on

2. Identify forces that could damage underpinnings

3. List potential threats

4. Compile individuals' judgments of the potential impact from each threat and determine if a consensus exists

5. Examine the overall threat pattern (How pervasive? Will it be sustained? What alternative futures can be constructed and planned for?)

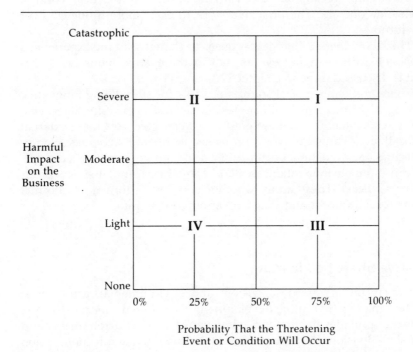

Probability That the Threatening
Event or Condition Will Occur

There are many other analytical tools, some more quantitative, others more qualitative. The key thing is to have a match between the type of targeted information and the means to interpret it.

Applying the Information All the monitoring information in the world will amount to nothing if we are unwilling or unable to use it for change. Sometimes the new information about change is more than we can take—or more than we are willing to confront. Overall, we tend to respond to signals for change in one of four ways:

1. *Inactive*—ignoring signals of change and failing to act

2. *Reactive*—ignoring the signals, not planning, and then taking crisis action when forced to

3. *Responsive*—noting the signals and planning actions accordingly

4. *Proactive*—anticipating the signals (even before they are clear to others), strategizing, and taking action to bring about a desired future

Organizations have predominant patterns in one of these four categories, too. The number of organizations in each category seems to increase from the "proactive" (the fewest) to the "inactive" (the most). From proactive to inactive, there is a continuum of willingness for managers and employees to face the truth about the real, practical world in which they operate. The proactive work to exert more influence over their future.

The key to these different responses to change is in the experienced *relationship* with changing events, not in the events themselves. This point is discussed more in chapter 12.

There are many special sources of ideas for stimulating innovative thinking. Conversations with professionals and peers, suppliers, customers, consultants, and new employees can give you new, external perspectives. Analysis of world problems, competitor activities, emerging technologies, customer complaints, and your organization's capabilities can stimulate innovations in all functional areas (not just new products or services). These apply whether you have a formal or informal system (or both) for monitoring or scanning to tap into.

Creative Strategic Planning

Your organization's purpose, vision, and environmental factors most clearly come together in the development of the organization's strategy. A strategy provides the first major link between your current reality and your shared vision. Purpose provides the fundamental direction, vision provides the destination, and strategy development provides the compass and road map. Together they constitute your organization's value-laden, fundamental attempt to *create a specific future.*

In its most basic terms, a strategy really *is* a statement of creativity: What future does your organization wish to create for itself? Strategy development must encompass what the people in your organization want to *be* as a group (your overall mission, purpose, business definition and short-term vision), what they want to *do* (your strategy and tactics), and what they want to *have* (your goals and achievements).

As discussed earlier in this chapter, the question of what your organization wants to *be* requires you to ask five questions:

1. What do *we* want to be?

2. What does the *market/society* want us to be?

3. What is it possible for *anyone* to be?

4. What do our capabilities lead us to be?

5. What are we called to be?

The answers to these five questions are usually found in an iterative process, typically starting with the initial question, "What do *we* want to be?", exploring the other questions in depth, and returning to this initial question. Many strategy development processes dwell too much on one of these questions as "the" key to success, when in fact the whole process will collapse if deliberation on any one is cut short.

There are two popular perspectives on strategy development: "strategic roadmapping" and "strategic positioning."

1. Strategic roadmapping typically: defines a goal or destination for the business and a map for getting there; assumes a future dominated by "certainties" (due to short time periods and/or stable environmental trends); employs single forecasts with $+/-$ variations; and emphasizes organizational structuring to implement the strategy with strong controls.

2. Strategic positioning typically: defines a direction for the future business and a compass for midcourse changes; assumes a future dominated more by "uncertainties" (due to longer time periods and/or more turbulent environmental developments); employs qualitatively different future scenarios; and emphasizes organizational capabilities and flexibility to implement the strategy opportunistically.

To many managers developing a strategy seems much easier than implementing it. The major flaw in many strategy-development processes is considering implementation the last step. In a very real way it is the *first* consideration.

Two factors are critical for successful implementation: decision makers must agree on and affirm a basic direction (alignment), and then they must commit themselves to the teamwork required to move in that direction (attunement). These qualities are usually difficult to elicit *after* a strategy has been developed; however, a well-designed strategy-development process can nurture them along the way.

The strategy development process is potentially the most vital way of establishing a culture of creative strategic management as both an operations ("line") and support staff function.

> Strategy development is a creative exercise . . . developing insights into your business and your markets, generating a wide variety of strategic options from which to choose, flushing out the best alternatives, and ensuring that plans are not only formulated but carried out.[5]

Most strategic planning processes are dominated by analysis—financial, market, competitor, technology, and so on—with relatively little attention paid to creativity. It is rare to even hear the words "creative" and "planning" in the same breath. How sad! Many elaborate plans become an end in themselves. They then sit on some manager's bookshelf until the next round of strategy development.

Consider, by contrast, musicians. They sometimes play with such connection—a communion, or attunement, if you will—that the performance rises above the composition itself. This is their "peak performance." A mediocre score cannot elicit an inspired performance. Yet, the strategies most organizations develop are composed as mediocre pieces, using an uninspired process and leading to uninspired implementation.

The APPEARE model is quite good for guiding the strategy-development process creatively.

1. Being "aware of your current situation" means doing the important financial, market, competitor, product, and operations analysis.

2. The "persistent vision" emerges in the iterative process of also "perceiving your alternatives," "entertaining your intuition," and "analyzing your alternatives."

3. Taking "realistic action" encompasses both the tactical planning and the actual implementation.

4. "Evaluating your results" completes the loop with feedback, to stimulate corrective actions by another round through APPEARE.

Given this perspective, here is an example of a highly effective and interactive approach to achieving the two major project objectives. It assumes the identification of two key participant groups from within your organization: (1) top executive staff ("EXEC."), who must have a broad vision of your organization's direction and are ultimately responsible for the strategy; and (2) key implementation staff ("IMPL."), who include those whose input is necessary for knowledgeable strategy development as well as those with key implementation roles. "Content" and "process" specialists from outside your organization can often contribute to and guide the strategy development.

Pages 168–169 map the process and key strategy-development questions being addressed. The chart identifies critical workshops only; important data gathering and discussions would, of course, occur between such workshops. The process requires collaboration between top executives and implementation staff in only three meetings, reflecting time and availability restrictions of many executives.

This process reflects the dual concerns of creating a compelling business strategy and a committed, capable organization to implement it. The final report would symbolize and document the accomplishment of both of these objectives. The process itself could well become a model for other major undertakings by the corporate planning group.

I mentioned earlier that alignment and attunement are critical to successful strategy implementation and that they are usually difficult to elicit *after* a strategy has been developed. In chapter 12 I discuss the issue of broad participation in more detail.

At the heart of the creative strategy-development process is an innovation search that encourages the creative freedom to generate options that are unique and relevant. The application of both linear and intuitive thinking can be an exciting, efficient, and effective approach to strategy development. For example, we once conducted a series of creative strategy workshops for a financial institution seeking an "umbrella" strategy for its Asia business. It had to incorporate local strategies for such diverse countries as Korea, Pakistan, and Australia. On day one of the first 3½-day workshop, we examined the current country-by-country demographics, market analyses, and so forth with the client. First-draft *country* strategies were formulated, giving us a baseline for looking at new ideas and alternative strategic themes.

Then, during the remainder of the workshop, the participants (forty people, including consultants, from seventeen countries) were given assignments such as the following:

- "You are no longer in charge of all products for a particular country. Instead, you are in charge of a single (named) product for all of Asia. What is your business development strategy?"

- "Use the design tree technique to develop alternative products and services and product-delivery methods starting with a given current product."

- "You are your major competitor. As this competitor, what is your strategy to wipe out all other competition in Asia?"

APPEARE STEP	PARTICIPANTS	KEY STRATEGY-DEVELOPMENT QUESTION(S) ADDRESSED
Workshop #1		
Persistent Vision	EXEC. and select IMPL.	Define project. Discuss "What do *we* want to be?" (initial variations), "What are we called to be?" and "What do we want to have?" (goals, achievements)
Workshop #2		
Persistent Vision	All IMPL.	Define and organize project. Discuss "What do *we* want to be?" and "What are we called to be?" (variations at lower levels in the hierarchy)
Workshop #3		
Aware of Situation, *Perceive Alternatives,* *Entertain Intuition*	Four IMPL. subgroups	Each conducts a brief Business Opportunity Search with focus areas based on key driving forces of alternate scenarios,* asking, "What does the *market* want us to be?" "What is it possible for *anyone/us* to be?" and "What would our capabilities lead us to be?"
Workshop #4		
Aware of Situation *Perceive Alternatives*	All IMPL.	Discuss opportunity ideas and integrate them for each scenario ("vertically" on the scenarios grid).**

---

* For example, your organization might have four qualitatively different scenarios based on driving forces of Global Economy, International Trade Avenues, Technology, and Social Values. The scenarios grid would be:

	SCENARIO			
DRIVING FORCES	**A**	**B**	**C**	**D**
Global economy	1a	1b	1c	1d
International trade avenues	2a	2b	2c	2d
Technology	3a	3b	3c	3d
Social values	4a	4b	4c	4d

Each group would select one of those driving forces and generate possible business opportunities *across* the scenarios grid (based on variations in that driving force).

** With a set of such possibilities for each scenario, the group would identify optional strategic themes—encompassing all four scenarios—for your organization. These themes would be elaborated on before and after the next EXEC./IMPL. session.

APPEARE STEP	PARTICIPANTS	KEY STRATEGY-DEVELOPMENT QUESTION(S) ADDRESSED
Workshop #4 (Cont'd)		
Entertain Intuition		Reexamine options for "What do *we* want to be?" and determine possible strategy *themes* (not specifics) across all scenarios—"What do we want to do?" in relation to "What do we want to have?"
Analyze Alternatives		
Workshop #5		
Entertain Intuition	EXEC. and select IMPL.	Reach consensus on "Therefore, what do we want to be?" and "What do we want to have?"
Analyze Alternatives		
Persistent Vision		
Workshop #6		
Realistic Action	All IMPL.	Translate strategic themes into more tangible options of strategic direction and positioning—"What do we want to do?" Tie these to capability assessments, "What is it possible for *us* to do?" (including an assessment of required resources and organizational capabilities)
Aware of Situation		
Workshop #7		
Realistic Action	EXEC. and select IMPL.	Reach consensus on strategy direction and positioning—"What do we want to do?"—in relation to "What do we want to be?" and "What do we want to have (as a result)?"
Workshop #8		
Realistic Action	IMPL.	Develop implementation tactics and action plans for the strategy—the next level of detail for "What do we want to do?"
Workshop #9		
Realistic Action	IMPL.	Review final report for broader communication and enrollment.

- "What is the nature of your business: financial, transaction processing, or what? What product and delivery strategies are most appropriate to each fundamental type of business?"

- "What is your notion of the ideal company for operating in your country? operating in Asia?"

- "Follow the instructions of a guided fantasy with your eyes closed." (Imagine taking a walk in a relaxing island setting and chatting with a wise, old person who advises, 'The key to your strategy is ————.' Find a box with an important symbol inside. Stroll through a town and find a new brochure announcing the company's new direction . . .)."

Each exercise provided more information and new insights, which began to fall into four major strategy themes. Following the first workshop, each country's participants developed four country strategies based on these themes. A second workshop began the integration process whereby an overall strategic imperative emerged. Initial financial targets could be formulated in light of the strategic imperative.

For projects such as these, pure "numbers crunching" and other demographic/analytical thinking would have missed ways to take advantage of many of the key opportunities identified for the Asian region. Intuition was needed to link together so many facts and diverse interests. Without it, there would have been no way to formulate and coordinate the investments in systems, promotions, and human resources needed to make the strategy operable.

Both linear and intuitive thinking are also needed to evaluate strategy alternatives and carry them out. Certainly evaluation would include *analyses* of competitor strengths and weaknesses, market demand, return on investment (ROI), investment needs and risk, and so on. *Intuition* plays an important role too, as data-based predictions reach their limit amidst the discontinuities of rapid technological and social change.

Along the way, expect some changes in your plan rather than a straight-line trip. As with the airliner flight discussed in chapter 4, success in travel is due to proper midcourse corrections. You can most easily know *why* and *when* to change if you have an adequate monitoring process. And you can most easily know *what* to change without losing sight of your goal if your original plan contains milestones, assigned roles, resource needs, and such.

Then, even when plans are thrown out and the whole organization is forced to align and atune itself to a new one quickly, you might at least agree with the skeptic who said, "Plans ain't worth a damn, but the planning process is invaluable."

IN CLOSING . . .

> ◖ *The thing I remember most about this one major project is putting the vision out, then allowing whatever happens to happen. The only thing common to everyone working on it was some sense of what was going to be the final result. For me to look back on my corporate experience, that element was present whenever something was done effectively in a group, though I wasn't aware of that at the time. I was aware of when I was trying to force my will on the group versus the times when I was synergetic and open to suggestions.*
>
> *Terry, political strategist* ◗

While there have been differing opinions about the causes, stability and longevity of the extraordinary growth of People Express Airlines, its president Donald Burr firmly believes in the power of a shared vision—a shared "ownership" that goes beyond the economic ownership that reinforces the employees' commitment:[6]

> [We live by] *six precepts . . . The first one, of course, is service, growth, and development of people. The second is to be the best provider of transportation of people. The third is to develop the best leadership. The fourth is to be a role model. The fifth is simplicity. The sixth is to maximize profits . . . If people ask, "Where are we going, where are we headed?" that's it.*
>
> *I tell everyone, "Make all the mistakes you want; fly the airplanes upside down. No problem. But just remember, we're always guided by those precepts: we take care of each other, and we take care of customers." Within those bounds, you can do just about anything.*
>
> *I really believe that we would never have been able to grow to $1 billion in annualized revenues in just under four years without our particular type of organization. If we had not put in the flat structure and the free environment, we never would have made it.*

To be sure, not all employees "buy" the precepts; some pilots call them "Kool-Aid." Yet Burr goes on to say,

> *It's really hard to articulate a vision so that people will get with it. My belief is that over time, the carpers will go somewhere else; they will become increasingly uncomfortable with the peer pressure from people who do get it. . . .*

Purpose, vision, and monitoring are all critical to a creative, practical strategy. To achieve excellence you need more than alignment behind someone else's purpose and vision. You need personal attunement with the purpose and vision: the purpose and vision must resonate with your sense of yourself. That is where your spirit, energy, and creativity for fulfilling the purpose and vision come from.

This is true at all levels, including new product introductions. Professor Modesto Maidique of Stanford studied why some product introductions succeed and why others fail.[7] He found it had little to do with technology and a great deal to do with internal collaboration and commitment. In fact, his top three recommendations for successful new-product introductions are:

1. Get a very clear understanding of the customer's business and needed cost/benefit ratios (the customer should be viewed as part of the organization).

2. Emphasize your own internal coordination.

3. Be sure top management is fully committed to the project.

This brings us back to the questions that began this chapter.

Q How *would* you and those you work with answer these questions:

Where are we going?

What do we stand for?

When are we most alive, purposeful, productive, and attuned to our business environment?

And what is *your* role in finding shared answers to these questions?

CHAPTER 11

"Institutionalizing" Innovation Where You Work

The Stop-and-Go of Ongoing Innovation

◖ *After many years of using my creativity to format computer printouts, manipulate data, and make worksheets, I was given the opportunity to break out of the numbers world into the world of people. I volunteered as leader of a yet untried project—to organize a career fair. I was asked to develop the idea, sell it to management, and then make it happen. This task proved to be the high point of my twelve years of working experience. The project took four months (I was holding down my regular job at the same time) and involved leading over 100 people. The experience charged my batteries and I was high for about three weeks after. I found I really enjoyed and was good at visualizing what an idea could become and then backing up to start creating it.*

Allison, former accountant in a computer company ◗

Have you ever been caught in heavy traffic when the next stoplight turns red just as yours turns green? Or perhaps worse, when all the stoplights were out and each intersection was an anxious battle for who goes first? What a blessing when you find a street where the lights are timed and you can breeze through at a steady speed.

As the needs of people and business change, there comes a time to install a new stopsign or light, to remodel an intersection, or perhaps to construct a bypass. At first, it can seem a nuisance to adapt to the changes—to remember to stop at the new stop sign or to bear with the interruptions of construction. In the long run, however, safety, convenience, and nerves are usually served well by the change.

In organizational life the stoplights are the policies and procedures, the reporting structures, the evaluation and reward systems, and the allocation of resources. (Too few resources, for example, result in too few roadways, which in turn results in traffic jams.) Ideally, these produce avenues of getting things done that are "timed streets." And when they have outgrown their usefulness and appropriateness—and if they themselves become the barriers to needed innovation—then comes the time to update them.

From an administrative standpoint the very nature of *innovating* can go against the grain of *producing.* Planning processes can be so management-dominated that the worthy ideas of nonmanagement personnel never "fit the plans." Short-term budget constraints can inhibit funding and motivation. Reward systems can reinforce efficiency to the detriment of risk taking. Segmented departments, which are often the most efficient for *delivering* a product, can become a barrier to *developing* new products.

How can you influence and manage the development of "timed streets"? The *first* step is to *understand* that:

a. the internal policies, structures, and such must be supportive of the organization's purpose and vision (form must be based on function);

b. organizational change has a definite "breathing" rhythm that must be taken into account (An organization is an organism, not a machine. It must breath, rejuvenate, and grow, or it dies.).

The *second* step is to be *willing and able to participate* in the process of "institutionalizing" innovation and change as a natural, exciting aspect of life at work. "Institutionalizing innovation" is not a contradiction in terms. It merely means developing systems that reinforce rather than inhibit strategic innovation in all parts of your organization.

The systems that most contribute to—or inhibit—the CREATIVE climate for innovation lie in two areas:

1. "Allegro Administration" (literally, "to minister to people quickly")

2. Evaluation Methods (how, when, and by what criteria ideas are evaluated)

This chapter can help you enhance your CREATIVE climate for innovation by focusing on "Allegro Administration" and "Evaluation Methods." These are the key systems that can stabilize, reinforce, and promote strategically appropriate innovation. These systems themselves can all be prime targets for improvement and ongoing innovation. The next chapter focuses on the process of managing the breathing rhythms—the transition periods for more updated innovation practices.

Each one of these topics merits volumes. The comments in these two chapters are aimed at the *context* for establishing stop-and-go lights that suit your organization's needs.

Allegro Administration

There are five components to allegro administration for innovation:

1. Innovation process (the steps an idea must go through to be implemented)
2. Budget, accounting, and resource-allocation methods
3. Performance appraisal/human-resource practices
4. Reporting structures
5. Information systems

Your Innovation Process

As you recall, the steps of our individual creative processes can be summarized as APPEARE:

A = Be AWARE of your complete current SITUATION

P = Be PERSISTENT in your VISION

P = PERCEIVE all your ALTERNATIVES

E = ENTERTAIN your INTUITIVE GUIDANCE

A = ASSESS your ALTERNATIVES

R = Be REALISTIC in your ACTIONS

E = EVALUATE your RESULTS

Sometimes we may follow a different order, perhaps with our intuitive guidance giving us new perceptions of alternatives.

The group innovation process includes all these steps as well. They may occur in a different order each time because the process of innovation is highly uncertain and unpredictable. James Brian Quinn, Dartmouth business professor, notes from thirty years of research on Bell Labs, GE, and other companies that "Few, if any, major innovations result from highly structured planning systems."

Parallel to this unpredictability, a new product idea must go through phases, key decision points, milestones, and functions to be implemented as part of a successful business. The Concomitance Model found on page 176 portrays this.[1]

You might think, "It's a wonder that any idea ever gets introduced or acted upon." Balancing the necessary stages in the evolution of an idea with the "disorganized" nature of the innovative process is the art of managing innovation.

You can examine your organization's process for innovation and determine:

THE CONCOMITANCE MODEL

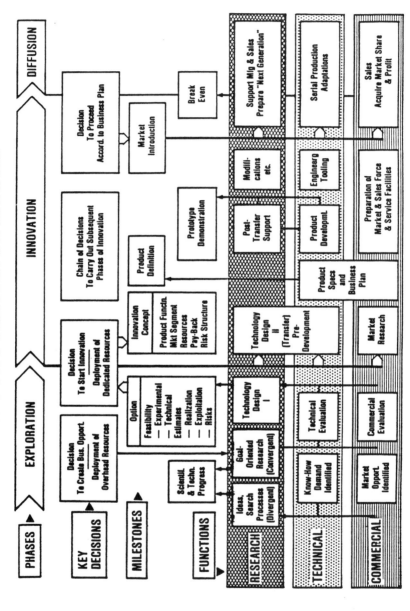

- Are there clear steps for submitting, reviewing, and screening ideas and providing feedback?

- Are the steps well understood by the people who are expected to submit ideas? (Do forms and procedures make the steps easy to follow?)

- Who is on the review committee(s), when does the committee meet, and so forth?

- Do the steps facilitate or inhibit the flow of ideas through the organization?

Sometimes a task force is called to chart a simple version of the organization's innovation process and issue a revised "roadmap." The consumer products company discussed in earlier chapters found the following needs for improving their innovation process.

- Institute a more formal process for identifying and evaluating ideas.

- Shorten the new-product cycle.

- Prevent premature killing of (good) ideas.

- Develop a system for transmitting ideas.

- Improve internal paperwork flow.

- Appreciate intuition more, as well as logic.

- Respond faster to changing market conditions.

- Reduce management approval time.

- Focus on long-term issues and problems.

- Give higher priority to ideas with very high potential return.

- Develop better criteria for judging ideas.

In response, management simplified their new-product process into three broad activities: ideation, development, and implementation:[2] A model of the new product process is found on page 178.

How well do you know the steps required to carry an idea through *your* organization? Is the process flexible or cumbersome? Is it simple enough to encourage you and others to invest your energy in an idea (for possible action)? Is it rigorous enough to hone and forge ideas into truly valuable concepts? If not, you and others need to aim your creative thinking at the process itself.

Perhaps the biggest problem in spurring innovation is connecting the idea people with the right sponsors. Many companies are experimenting with new ways to make these connections *outside* of the official

MODEL OF NEW-PRODUCT PROCESS

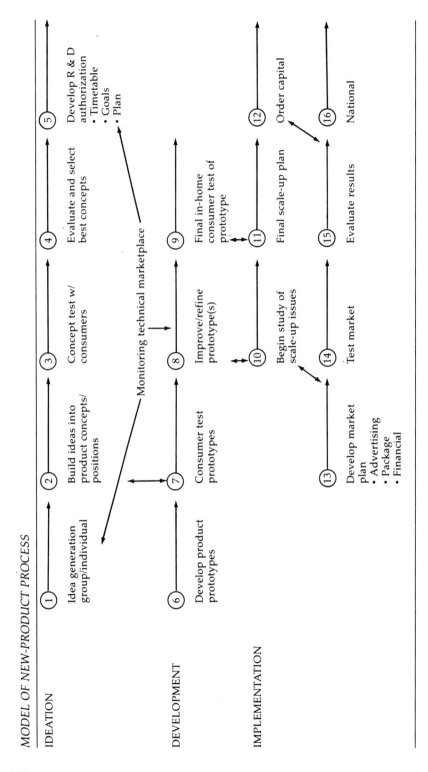

innovation process. In a sense, they are attempts to "formalize" support for unpredictable, "skunkworks" innovation. Of particular popularity are: (1) innovation facilitators and committees, (2) innovation fairs, and (3) innovation centers.

Innovation Facilitators and Committees Companies like Kodak and Xerox have developed alternate routes for their innovation process to supplement the normal product development channels. At Kodak, the normal stages of the innovation process look like this:[3]

STAGES IN THE INNOVATION PROCESS

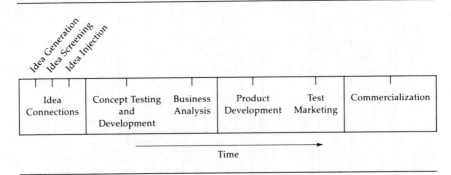

Kodak has instituted the role of "innovation facilitator" to help make the connection between idea people (those who generate ideas not related to their everyday job) and potential sponsors (those who have the necessary resources to help get the idea off the ground). Organized as the Office of Innovation Network (OIN), their program operates under a person-oriented, not idea-oriented philosophy: nurture people and develop their strengths; ideas will come about naturally. In other words, people are *not* just "baggage" that sometimes come attached to the ("important") good ideas.

With OIN, when a person submits an idea, a facilitator contacts appropriate "consultants"—knowledgeable specialists—within Kodak to help formulate and possibly revise the idea. A meeting (or meetings) is held to test and revise the idea further. All the while, the idea is never "taken over" by the facilitator. The idea person must remain the idea champion; the facilitator "merely" makes the connections and guides the process. Worthy ideas are ultimately connected to sponsors for a seed grant (perhaps $25,000) to develop the idea through a core team of interested personnel.

The steps in this process are described by the "process" map found on page 180.[4]

About 80 percent of the time the sponsors are in current operating divisions of Kodak (but different units from the idea champion's report-

INNOVATION / IDEA EVALUATION SYSTEM*

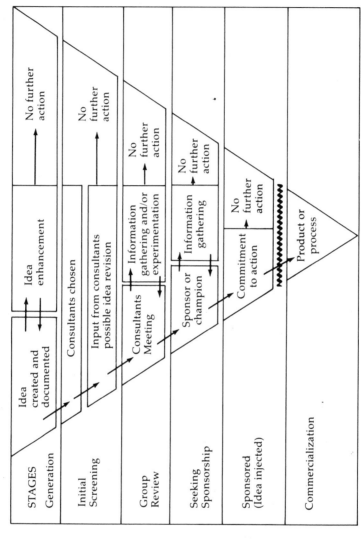

STAGES		
Generation	Idea created and documented	Idea enhancement → No further action
Initial Screening	Consultants chosen	No further action
	Input from consultants possible idea revision	
Group Review	Consultants Meeting	Information gathering and/or experimentation → No further action
Seeking Sponsorship	Sponsor or champion	Information gathering → No further action
Sponsored (Idea injected)	Commitment to action	No further action
Commercialization	Product or process	

* Office of Innovation, Kodak Research Laboratories

ing line). The 20 percent that represent new business areas for Kodak go to a New Opportunity Development (NOD) group that operates as an internal "venture capital" unit. NOD evaluates ideas submitted to it for sponsorship by four initial criteria:

1. Can it be done (technically)?

2. What does Kodak bring to the party (e.g., capability and strategic fit)?

3. Is it big enough as a potential business?

4. What might be the return on invested resources (big enough benefits/nuisance ratio)?

If the idea is promising, the idea champion must recruit a management team—where the "general manager" might not necessarily be the idea champion—with the possibility of getting a five-year funding commitment. The management team must then create a full-fledged business plan for approval. Ventures such as Lambdek Fiber Optics, Beta Physics, & Eastman Communications were formed by this process.

This process itself was an innovation spurred by a product champion, Robert Rosenfeld, back in 1979. Starting with one Office of Innovation in a divisional laboratory, Kodak has recently made the commitment to about twenty such OI's worldwide. Collectively they are referred to as the Office of Innovation Network. The market success of products stemming from this alternative innovation process has warranted such commitment.

The position of innovation facilitator is an essential ingredient to this success. The facilitators need to have a balance of skills in:

- getting commitment (from organizations, consultants and sponsors),
- running group meetings,
- guiding business planning,
- showing genuine caring for people and ideas,
- communicating with a wide variety of personalities,
- understanding technical subjects,
- thinking on their feet,
- networking,
- coping with multiple tasks and projects (while remaining fresh and genuine with the next person with a "great idea").

The rewards for participating in the OIN program have not been financially oriented. The primary motivations of idea contributors have been "getting my ideas heard" and "seeing my ideas acted on." Financial incentives, although not necessary to launch the program, are nevertheless being seen as an area for improvement if the program is to continue its past success.

Xerox has another example of an alternative process for developing innovation ideas. In 1980 the Reprographic Business Group founded the Innovation Opportunity Program (IOP) to administer funds for exploratory technology and product projects. The IOP team's responsibilities include the following:

a. Assisting idea originators in developing their proposals

b. Providing recognition for the efforts of innovators

c. Promoting an environment that inspires innovation

d. Communicating the goals of the group to *all* employees

The steps of the process are simple. Once you have an idea and contact an IOP team member, a back-up person is selected who helps you prepare a brief informal presentation of your idea to the IOP team. The IOP team listens, ask questions, and then meets privately to rate the merits of your idea based on (1) your initiative and enthusiasm, (2) the effect on the environment, (3) the value to Xerox, (4) the value to society, (5) the resources required, and (6) the probability of success. IOP then provides resources to develop the idea until it can be delivered to an appropriate organization for further development. One example of an idea that made it this way, but not through the normal channels, was a special integrated circuit connector for the Xerox 1075 that dramatically reduced spare parts costs. Another was the development of a color printer despite severe resource constraints.

Robert Gundlach, a senior research fellow, comments,[5]

> *The essential ingredients . . . are unwavering support by management (emotional commitment as well as funding of $\frac{1}{2}$ to 1 million per year . . .); a dedicated, but democratic manager of the program; and a council of 6–12 non-managers recognized by their peers as innovators . . . Finally I would stress the primary purpose should be to convey a positive attitude of receptivity to new ideas, and secondly the implementation of these ideas into products. Putting the major emphasis on the improved environment for change conveys the important implication of faith that if we meet the first goals successfully, the second will be realized automatically.*

Innovators Fairs Another avenue for matching people's ideas with resources is the "innovators fair" held by 3M, Xerox, Kodak, and others. Often they are the result of a committed product champion spurring the innovation climate. At Owens-Corning, for example, Frank Burroughs and his cohorts in Ohio went beyond executive skepticism and produced a local innovation fair that ultimately had executives and technical people flying in from around the country. They advertised for people to present their ideas informally at tables in a conference room, expecting twenty to twenty-five presenters to sign up. They ended up using meeting rooms and hallways to accommodate over seventy-five presenters. Enthusiastic linkages were made even with executives who normally follow very rigid, formal evaluation schemes.

Innovation Centers Hallmark Inc. has developed a new Technology and Innovation Center with a building specifically designed to foster creativity among artists, craftsmen, technicians, and scientists.[6] As the intended focus of the company's new product development, it is an outgrowth of a creative workshop that gives people an environment to let ideas run wild. Both the center and the workshop get artists closer to technical and manufacturing capabilities that can stimulate new product ideas.

Budget and Accounting Methods

Budget and accounting systems are designed to monitor and control the "exchange of value" from each reporting unit. Specific budget allocations are fought for vigorously, but the whole *process* of budgeting is an even larger issue.

Although your budgeting and control process may be intended to support your organization's purpose and vision, often those systems themselves can become the show. They end up replacing the purpose and vision as the prime motivator and measure of manager or unit success. But meeting the budget does not necessarily mean meeting the goals of your organization!

Budget systems *can* be employed to stimulate innovation rather than just measure and control operations. Many companies have reorganized as a "free enterprise" of departments, a structure that demands a budget- and performance-appraisal system to support it. On the other hand, some organizations focus so heavily on short-term profits, return on investment (or return on assets), and revenues that long-term purpose, positioning, and innovation are lost.

In one case an electronics manufacturer had product development and sales in two different "profit center" accounting structures. The sales group's costing for taking a new product to market was appropriate for established products but overwhelmed the potential profitability of start-up product lines. Innovation was severely stifled by these accounting procedures.

The budget system must provide alternate funds for ideas being developed outside the normal lines of authority. Texas Instruments, for example, instituted a senior management committee to review such ideas, but with special ground rules. The employee only had to get one of the twelve members on the committee to sponsor his or her idea to be able to appear before the committee. The committee had the authority to appropriate funding for a thorough initial study of the idea. Before committing funds, an advisory panel of technical or marketing people reviewed the idea and a project plan was submitted with personnel, budget, time, and other resource requirements.

The budgeting process can also reward those who evoke new ideas for staying ahead of a changing external environment. A "consulting" leadership style at budget time can highlight investments in innovative ideas much better than a top-down, "tell" or "sell" style.

In summary,

Q Does your budget and accounting system support the show or run the show?

Does the planning and budgeting process elicit ideas for innovations to meet environmental demands?

Are budget allocations decided from the top down or through a consultative and participative process?

Are there alternate routes for funding innovative ideas?

Performance Appraisal and Human Resource Practices

The bottom line of a clear and effective innovation process is profit, productivity, and return on investment not only in a *business* sense but in *human* terms as well. As the people prosper by effective channeling of creative talent, so does the organization.

In human terms "profit" becomes *integrity and trust*—people becoming whole and trusting that giving their best will be rewarded with fairness and respect. "Productivity" becomes *commitment and competence*—employees being all they can be, in support of the shared vision. "Return on investment" becomes increased *teamwork, patience and initiative*—people working together for long-term organizational and personal health.

An organization's long-term business profit, productivity, and return on investment *do* stem from the prosperity of its people, as well as the other way around. Investment and originality in the innovation process and human resource practices can be the most significant factor in long-term organizational health.

In one research study sixty-five vice-presidents of human resources in major companies identified the corporations they thought had been the most progressive in their human resource systems and practices. Forty-seven companies were identified, including IBM and General Electric. A non-nominated firm in the same industry and with similar assets, sales, and number of employees was paired with each progressive company. The twenty-year financial performances of each pair were compared, and "the companies with reputations for progressive human-resource practices were significantly higher in long-term profitability and financial growth than their counterparts."[7]

Innovations in human resource practices and organizational designs, therefore, might be considered "innovation-producing innovations." When you have instituted a CREATIVE climate for innovation, your organization can produce appropriate innovations when and where needed. You and others are better able to generate new solutions, work collaboratively to implement them, and experience rewards for your contributions.

Many organizations say they want innovation, but when it comes down to *what* their managers are really held accountable for—as reflected in their performance appraisals and financial accountabilities—innovation is rarely mentioned in specific terms. The 3M Company, as one exception, is well known for its requirement that a division earn 25 percent of its revenues from products introduced within the previous five years; this demands constant product innovation. They support that objective at the individual level with both financial and *time* resources. As their professionals go about their work, they are encouraged to spend up to 15 percent of their time working on dreams of their own, thereby uncovering enough nuggets to convince their bosses to stake out a claim.

In performance appraisal the real questions are, "How would you recognize creativity if you had it?" and "What type of recognition do you want to give for it?"

It might seem impossible to set goals for creativity and innovation. And there are two ways we can set such goals:

1. Directly for specific creative output, such as "Develop two new product concepts that meet the following criteria" or "Produce a new information-management system to meet these requirements."

2. Indirectly for realistic "stretch" goals that require creativity to be fulfilled, such as "Achieve a 25 percent increase in sales revenues with current staff" or "Maintain the productivity output of this group with 10 percent decrease in budget."

Within these goals, we can evaluate how well a person or group:

- Fosters an open, questioning environment for new ideas

- Develops new ideas worth testing

- Produces innovative results

- Accomplishes positive results using documented creative approaches

In both cases you then praise the person or group not only for achieving the tangible results you wanted, but also for the quality of original thinking that helped them achieve those results. In the consumer products company, for example, one thing that managers soon began listening for was the use of the words "creative" and "innovative" in the description of a new product or marketing idea. The idea may or may not have been adopted, but the person was encouraged to continue looking for new and creative approaches to their work.

As we discussed in chapter 4, there is reluctance within most of us toward having our performance measured; yet without appraisals we can't get acknowledgement and recognition for our successes. No performance appraisal methodology will work for all people, in all professions, in all situations. The guidelines that best foster creativity build on four widespread principles that are often difficult to implement:

1. Individuals must be involved in setting and evaluating their performance

2. Both formal and informal feedback are required

3. Flexibility in evaluating results is essential

4. Risk and failure need a degree of nonpunitive tolerance (even encouragement!)

In chapter 3, we discussed several of the many different rewards that might be appropriate, including promotions, special compensation, recognition, and the opportunity to pursue an idea. Sometimes opportunity and autonomy are sufficient. A dual career ladder for managers and technicians within an organization is one example of this. IBM's Fellows program, giving scientists up to five years to work on a project of their own choosing, is another. Harry Shaw, president of the Huffy Corporation, tells the story of one ingenious inventor who developed a new bicycle that was used to win six Olympic medals. As a reward for his "intrapreneuring" efforts he was put in charge of manufacturing the bicycle for the public, but his talents were in inventing, not managing. He was replaced—a discouraging end to a success story. It points out the need for dual career systems, where people can follow career progressions suitable to their real talents. Such dual career systems are often implemented—at IBM, 3M, and Du Pont, for example—but, unfortunately, most cultures view the technical (nonmanagerial) paths as lower in status.

Similarly, there is a need for dual managerial paths—one for the entrepreneurial "starters" and one for the stabilizing "sustainers." This can help enhance the motivation and contribution of key managerial talent.

It's up to you and your creativity to develop your own menu of rewards. Innovation needs a safe place to risk and be creative. That means not using the performance appraisal process as a "punitive motivator." Les Krough, a general manager of 3M's New Business Ventures Division back in 1969 stated this quite well:[8]

> *A man who can do a good job with our existing business is an important cog in the whole establishment. But you have to stimulate people to move out of that by clearly stating that growth is the corporate objective. He has to have before him continually examples of people who have been rewarded for taking the risks inherent in getting us into new business. But more than this, risk-takers have to have some sort of security. If his project fails, for no fault of his own, you've got to have a place for him to go.*

Reporting Structures

As you have seen, there are many approaches to fostering a creative climate for innovation that don't require changing your reporting structures. Certainly there are also times when it helps to reorganize or introduce new business units.

In one health insurance company, the senior vice-president of finance took over the product-development function—a non-obvious assignment for a finance executive—because he would be able to provide a fresh outlook on how better to meet customer needs. Decentralization was used by another firm, Allstate Insurance, to revitalize. President Richard Haayen, who describes the firm as "50 years old and a victim of middle-age spread," says an entire middle-management tier has been eliminated. As one example of positive results from this decentralization, employees in some states have devised more competitive pricing systems.

In 1981, 3M's chairman, Lewis Lehr, concluded that changes were needed in the very foundation of 3M's organizational structure. Long before it became fashionable, 3M had always broken itself up into small business units, with three to four times as many divisions as most companies its size. Each division worked autonomously, doing its own research, manufacturing, and marketing. But based on competitive factors and Lehr's vision, another step was needed. To foster a further sense of pride and ownership in ideas, the divisions created even smaller companies specifically for developing new products. Whole teams in these business-development units are responsible for seeing a product through from development to introduction. The team is evaluated as a whole; if marketing screws up, that's the engineer's problem too. 3M's goal is to produce products that meet customer needs quickly and more accurately.

Many companies are setting up "internal ventures" in which a project team is put to a task outside of the normal reporting structures, or even outside of the normal compensation guidelines. These "intrapreneur" structures have advantages, particularly when there's a need to respond to market and technological changes as quickly as smaller competitive companies can. IBM's PC was developed this way. All of their venture teams are headed by what they call "true believers." Internal ventures are also often instituted when the normal structures have simply become too unwieldy.

Neither reorganization nor internal venturing is the cure-all for stimulating successful innovations. They may often promote disruption rather than innovation. Reporting structures share a common quality with the other major systems (innovation process, budget and accounting, and performance appraisal): if well conceived, well planned, and well implemented, changes in these systems can make a significant difference in innovativeness for the short term. For changes in these systems to be effective for long-term innovation, however, the other elements of the CREATIVE climate for innovation must also be addressed.

An overall principle to guide that process is the promotion of what Rosabeth Moss Kanter calls "integrative" rather than "segmentalist" interactions among people.[9] As she puts it, integrativeness is "the willingness to move beyond received wisdom, to combine ideas from unconnected sources, to embrace change as an opportunity to test limits." Segmentalism is "concerned with compartmentalizing actions, events, and problems and keeping each piece isolated from the others."

Integrative thinking is necessary for innovation and is more likely found where there are integrative structures: few boundaries between units, mechanisms to exchange information, multiple functions involved in decisions, and so forth. Segmentalist structures, which reinforce anticreative thinking, have a large number of units walled off from each other; problems are carved into pieces for specialists who work in isolation. "Even innovation itself can become a specialty in segmentalist systems—something given to the R & D department to take care of so that no one else has to worry about it."

Segmentalist organizations also tend to value the logical, linear modes of problem solving to the exclusion of the intuitive. The integrative organization values fluency in both modes.

There's an inertia in maintaining the status quo; yet there's also a momentum in continuing to grow, live, and thrive. Changes in allegro administration can sometimes be perceived as a threat to personal or group survival in the organization. But change is *not* the real threat to survival: the *lack* of change is! You always have the opportunity to promote the growing, thriving organization rather than the unchanging, declining one.

Information Systems

We saw in chapter 10 that the greatest stimulator to innovation can be information. Whether or not information is effectively collected, analyzed and distributed can make or break the innovative efforts of your organization. This is true not only with externally generated information, but also with information from internal sources.

More and more we use information systems (broadly defined as the combined efforts of computers, telephones, and many other devices to exchange information) to help us collect, store, process, communicate and display information. Often these information systems are themselves innovations. The SABRE reservation system of American Airlines or the HOLIDEX reservation system of Holiday Inns are two such innovations. Information systems are also the midwives for innovations in marketing, engineering, sales and many other areas.

As part of the CREATIVE climate for innovation, information systems can play a critical role in the stimulation, flow and implementation of ideas. For example, automakers are using computer-aided design (CAD) workstations to model and test concepts without the expense and time involved in making three-dimensional prototypes. Grocery stores are issuing their own credit cards not simply to make purchasing easier for their customers, but to track customer purchasing patterns and to develop new marketing strategies.

Information systems can personally empower us to respond more creatively to change by increasing our ability to stimulate and implement new ideas and to solve problems. Anyone who has composed or corrected a letter or report with the help of a word processor knows how easily new and creative combinations of words, tables and graphics can be created. Taken in the right spirit, computers can help us as SPIRITED people to exercise more freely our inventiveness, our persistence, and even our spontaneity with new ideas. Personal computers can offer assistance during each step of the APPEARE process. "Brainstorming" and "outlining" programs such as MaxThink and ThinkTank can help us to generate ideas and see alternatives. Graphics packages such as MacPaint can help us conceptualize and symbolize important steps to defining problems as well as presenting solutions. Programs for financial spread sheets and for project planning can help us understand our situations, analyze alternatives and take realistic action. Networks of computer users, such as those connected by electronic mail communications, can often get a wide variety of responses to a call for more ideas in a short period of time.

At the other end of the computer spectrum, mainframe software systems can make a significant difference in whether an organization has a Creative Edge or not. As a case in point, consider a particular bank that had no central customer information files. A customer (or a husband-and-wife household) might have seven accounts (checking, savings, mortgage, credit card, certificate-of-deposit, car loan, and teller-machine card) but the bank had no way of bringing the records of these accounts together. If, for example, the customer wanted to change his/her address, it took seven different transactions to do it, an expensive administrative procedure that became more complicated if there were slight variations in the way the customer's name was listed on each account.

By developing a new cross-reference information system, the bank not only solved this administrative problem but also created an essential tool for new product development. At one point, the new product staff wanted to increase the number of checking accounts without increasing the number of tellers needed to service the customers. They used their cross-reference systems to help profile those people who were high users of automatic teller machines (ATMs), and eventually established a new service to get people to switch their checking accounts from competitor banks.

Also with the cross-reference system, marketing profiles demonstrated trends such as "New checking account customers are most likely to switch their other credit accounts from other banks within the first 60 days; after that, the likelihood drops off significantly." Using this information, the new service offered a very low monthly fee with unlimited checking, but with a fee for going to a teller for a transaction they could have done by the ATM. The very low fee was based in part on being able to use the cross-reference system for selling other services and accounts to the new customers.

Not having such accessible information can be an insurmountable barrier to creative product development, especially in industries changing as fast as the financial industry. For example, some insurance companies sell exclusively through agents who are not employees. These agents have the most frequent direct feedback from customers about the products and service of the insurance carriers. If an insurance company has no distributed information systems linking it to these agents, then the company's only access to customer information is by paper and by word-of-mouth, filtered through those in charge of agent-liaison. This situation can make the development of new products and new means of servicing customers much less responsive and less "on target" with customer needs.

The use of information systems also raises many questions for other issues of the CREATIVE climate for innovation, including:

- Environmental Monitoring: Will information get wider distribution to where problems exist "locally?" Will information about the business environment get more attention and generate faster responses?

- Reporting Structure: Does a wider use of personal computers (plus wider access to information) mean more decentralization or a flatter pyramid? How will information systems alter the jobs and careers of middle managers?

- Collaboration and Roles: Will a wider distribution of information processing mean more or less balance of power in departmental collaborations? Will "Lone Rangers" make decisions without collaborating by assuming that they have all the information they need?

- Innovation Process and Evaluation Methods: Will the steps for problem solving—perceive the situation, define the problem, analyze, generate alternatives, etc.—proceed faster and more effectively, yet with broader participation? Will the quantitative and qualitative criteria be applied appropriately at different evaluation stages?

- Balance of Intuition and Logic: Will people develop fluency in intuition as well as logic, or will they rely on data-crunching to drive their own decision-making? Will "what if" games enhance balanced decision-making?

As these questions suggest, information systems could be the greatest reinforcement—or the greatest hindrance—to our other efforts to implement a CREATIVE climate.

Information systems are often justified and designed for their impact on productivity. The savings in costs, time, and effort can generate significant competitive advantages. However, their impact on innovation can be just as significant. Many organizations suffer from a missed opportunity in not making creativity an equally-urgent justification and criteria for developing information systems. Our worklives are filled with more and more information, coming to us faster and faster. How we employ information systems can dramatically affect how we employ our creativity, and ultimately on how we enhance our employment.

Evaluation Processes

Innovation at large companies decreases as the implicit or explicit evaluation process becomes burdensome. In two better-known examples,

Chester Carlson searched fifteen years for a corporate sponsor of his new "xerography" process; and Charles Hall's original aluminum process was spurned because the material was "only usable for trinkets and jewelry." On the other hand, for all the ideas that deserved to be nurtured and weren't, there may be just as many that deserved to be killed and weren't.

One chemical company estimates that for every 300 products from their R & D labs, 3 make it to the marketplace and 1 makes it big. It is more typical across industries that for every 100 products, 25 get analyzed seriously, 10 go into product development, 4 get market tested, 2 get marketed, and 1 gets added to the product line.

Everyone in an organization has a role in making the evaluation process work effectively. You can develop new ways to optimize the evaluation system by considering:

- *Why evaluate?* Is it to decide funding, congruence with organizational goals, success, graduation to a new phase of idea development, comparison with competing ideas in the market, or what?

- *What is evaluated?* Is there a process for ideas other than new products, including new work methods, new marketing concepts, new reporting structures, and so on?

- *Who evaluates at each stage?* What participation from different levels and functions is appropriate and how is it encouraged? Does everyone in the organization know the system and how it works?

- *When is evaluation conducted?* How many different stages and separate reviews are there in the life of a project?

- *How is the evaluation conducted?* What methods of evluation are used? What are the criteria at each stage of the process? How are quantitative measures integrated with the more subjective, intuitive norms often used by top management? Are both external attractiveness and internal fit considered?

- *Where are rejected ideas stored for later consideration?* How does the evaluation system call attention to new priorities for innovation?

Sometimes there is a mismatch of evaluation methods and the stage of development. The typical pattern of the decision points involved in new-product development and evaluation is found on page 193.

Too many criteria or excessive quantification too early can kill promising ideas or discourage new ones from even being posed. Too much quantification can also lead to the "creation" of data merely to satisfy the system. Conversely, waiting until late in a project's life to apply key criteria can waste valuable resources.

In the consumer products company discussed earlier, management

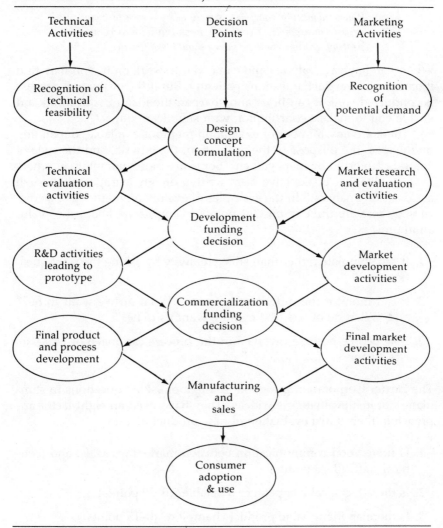

discovered that the presence of a four-page form with twenty to twenty-five criteria discouraged the submission of new product ideas and encouraged "hollow" market projection data. They instituted a new procedure saying,

> *We as managers need to ask 4 questions early in a project:*
>
> • *Does this fit with corporate strategy?*
>
> • *Does someone (consumers) want this?*
>
> • *Is the technology likely to be available?*

• *Do we have the people to champion it and do it?*

When submitting your idea, simply address each of these issues. For example, if your idea doesn't fit with corporate strategy, tell us why the strategy should be different.

When you have a great idea and can't wait to work on it, you may find it hard to answer such questions patiently. But if the evaluation process is conducted well, it can hone and sharpen the idea in your mind and enable you to make a contribution with significant impact.

There are any number of evaluation methods available, depending on the idea, the purpose of the evaluation, the criteria, and the style of analysis preferred by management. Some are more quantitative, others more qualitative. Books have been written on single approaches, such as "decision analysis." In the examples of evaluation methods to be discussed there are three important points to emphasize about the evaluation process:

1. It can stimulate rather than stifle creativity by asking the right questions.

2. It can monitor the difference between "What do we want to be?" and "What do others (the market) want us to be?"

3. *How* people become involved in the process can make or break its effectiveness.

The Dexter Corporation is a good example of asking questions to stimulate, not just evaluate, new ideas. They have used an eight-factor approach in their initial evaluation of new product ideas:

1. Is there good communication between marketing, sales, and technical staff? (0–20 points)

2. Is there technical competence available? (0–20 points)

3. Is there an identifiable product champion? (0–15 points)

4. Are there good market opportunities? (0–15 points)

5. Are there good applications of our technical strengths? (0–10 points)

6. Is there top-management interest? (0–10 points)

7. Are there good competitive advantages? (0–5 points)

8. How good is the timing? (0–5 points)

Notice that 65 of the 100 possible points are for organization and personnel considerations. And even in the market and technology criteria, qualitative judgment is mixed with possible quantitative measures.

In another case, Udell and Baker at the University of Oregon have developed one well-known instrument for evaluating new-product innovations that includes thirty-three criteria in five general factors:

1. Societal factors—such as legality and safety

2. Business risk factors—such as investment costs and payback period

3. Demand analysis factors—such as potential sales and trend of demand

4. Market acceptance factors—such as promotion costs and visibility of advantages

5. Competitive factors—such as appearance, durability, and price

A third example involved an industrial products group of a Fortune 500 company. To determine a new match between "What do we want to be?" and "What might the market want us to be?" we conducted an Innovation Search to identify possible new business concepts for the next five to ten years. The top concepts were then evaluated through a value-indexing process.

A value-indexing process is an alternative to a strict, quantitative approach to decision making. It gets people involved in choosing the opportunities that best fit the purpose and vision of the organization. The entire evaluation process, not just the criteria themselves, can complete the learning cycle that forges group cohesion and the commitment to make a difference *together*.

A set of sixteen criteria were defined, divided into two categories: (1) eight "external attractiveness" factors (those related to markets, competition, technology, etc.) and (2) eight "internal fit" factors (those related to corporate capabilities, business purpose, etc.). External factors are usually less under the organization's control than internal factors.

Statements were developed describing relative levels of meeting each criterion factor. For example, "potential growth rate" had levels similar to: 0–5 percent, 5–9 percent, 10–12 percent, 13–15 percent, 16 percent or more. Each statement was then assigned an index value on a 1–10 scale, such as:

STATEMENT	INDEX VALUE
16 + %	10
13–15%	8
10–12%	5
5–9%	2
0–5%	0

Data were gathered on the top twelve business concepts from the Innovation Search. The data were used to see how each concept scored on the 1–10 scales for each criterion factor. For example, if the estimated growth potential was 14 percent, a score of eight was given for that criterion. These determinations required extensive discussion among the members, a key part of the process.

At the end, scores were weighted if some criteria were more important. The scores were added for all "external attractiveness" criteria and for all "internal fit" criteria. For example, out of a possible score of 80 for external attractiveness (8 factors × 10 maximum points each), a concept might have scored a 50; the score for internal fit might have been 60 out of a possible 80. The scores for each concept were plotted on a matrix as shown below.

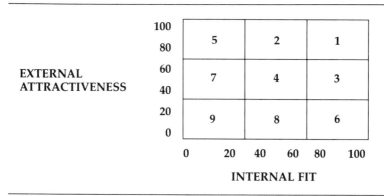

The example with scores of (50, 65) therefore would have fallen in section 4. This matrix can show how a concept may be very attractive but a poor fit (quadrant 5), or perhaps a very good fit but very unattractive (quadrant 6). It is usually easier to take the former and modify the internal fit (move from a 5 to a 2 or 1) because internal factors are more under management's control. It is more difficult to alter market attractiveness factors (going from a 6 to a 3 or 1). Therefore, the business opportunities would generally receive a priority for development depending on the square each fell into: from 1 to 9 as shown above.

The discussions surrounding the interpretation of this matrix and the decisions about which specific business opportunities to pursue were important to the group's alignment and attunement. Even more, a purpose and vision emerged that redefined the overriding nature of their business.

The key to an effective evaluation system is what works for the people involved. This means involvement and safety. At Owens-Corning, their venture screening process is coordinated by project personnel, not management or staff. In this way they hope the project personnel can increase their influence on and acceptance of the evaluation process. Regarding a "safe" environment, an often quoted ad by 3M says it all: "3M has made a lot of mistakes. We're very proud of some of them."[10]

IN CLOSING . . .

▶ *To provide new funding for projects, we found a way to furnish our labs and get some research money through companies that were making the products. When they were testing a new product, instead of testing it in a lab, they would bring it over to us. We got a lot of things on trial. And sometimes we test a drug, so we'll get money for part of the research.*

In another area, management was considering having benefits that were interchangeable, based on an employee survey. For example, they pay 50 percent of our children's tuition of any college they go to. Rather than saying everybody had that— *when some people don't have children—they now say, "OK, this person may want more life insurance, dental insurance, or eye care insurance." So they are supplying us with a basic core of medical benefits, but above that we can choose whatever we want up to a certain amount.*

They're also doing the interviews to find out how people feel about their jobs and their employer, trying to see how maybe they can change different relationships or how they get raises. The whole idea is to make people feel like they're a part. It's worked in some cases—it's done better than not doing anything at all. At least there's an effort.

Trena, manager in a hospital cardiac surgery unit ▶

Allegro administration and the evaluation process can provide the structure for ongoing productivity *and* the basis for ongoing innovation. With the proper structures an ongoing CREATIVE climate for innovation will have the best chance for long-term (and short-term) results.

Relying solely on structural changes to promote innovation will produce only short-term benefits for our organizations. In fact, structural change may not produce any benefit at all, given the disruption that may occur. They *support* life in the body of the organization, but they don't *give* the life. Our spiritedness and inspiration give that life.

When the structures themselves become the goal and measure of your work life ("meeting budget," "fitting in," etc.), you can be sure that you and your organization are heading for a state of burnout.

Your purpose and vision are your inspiration and passion. Let them guide you in designing the stop-and-go lights of ongoing innovation. Let them also guide your collaboration, roles and rewards, monitoring of the environment, change management, intuition and logic, and evaluation methods.

Systems and structures are the instruments, not the music. *We* are the composers, playing the melodies and harmonies of purpose and vision. We need not be limited by poor instruments—instruments that dictate the melody.

For each of us the music comes from within. For our organizations the music comes from the alignment and attunement of all who work there.

Can you hear it? The thoughts of Kabir can help us listen more closely.[11]

> *The flute of the Infinite is played without ceasing, and its*
>   *sound is love:*
> *When love renounces all limits, it reaches truth.*
> *How widely the fragrance spreads! It has no end, nothing*
>   *stands in its way.*
> *The form of this melody is bright like a million suns.*

CHAPTER 12

Rejuvenating Your Workplace through Planned Change

Sir Thrival and the Change Dragon

📧 *Today I facilitated a meeting with the head of clinic nursing and her seven nursing supervisors. This is intended to be a whole series of meetings that I've worked out with the nursing director. The purpose of the meetings is to start moving these nursing supervisors towards becoming real managers rather than "hands-on" lead nurses as it were.*

We opened up by talking about the purpose of these meetings and then about their reservations and concerns. These concerns took at least fifty minutes instead of the twenty I planned. It was time enormously well spent. I could actually see a difference in the room from when we started discussing the purpose of the meeting. They had been sitting there, scared. Some specific fears were responded to in the moment by the clinic director. In other cases it was simply acknowledged that this was something we take into account while going through this process.

What we did, essentially, was legitimize the fears—and just listen to them. It just does wonders when you do that . . . individually and in groups.

Tom, organization consultant for an accounting firm 🖐

Once there were two knights, Sir Vival and Sir Thrival.[1] Sir Vival loved to fight. He saw all of life in terms of competition. He fought every situation, every person, and every event that didn't fit into his picture of the ideal life. Anything that dared ask him to adjust to it was seen as a Change Dragon to be slain. Because Change was always upsetting his ideal life, he was always doing battle; and when his Change Dragons were asleep, he was too tired to do anything but sleep also. His life became a matter of constant battle, and he grew weary. Mere survival became his ultimate victory.

Sir Thrival also loved competition, but he could distinguish between fantasy opponents and true challenges. His Change Dragons were fewer, and he could devote more time to anticipating their next moves. He enjoyed the sport itself, but he recognized that slaying the dragons would bring an end to his sport. His competitiveness was a type of collaboration, for without each other there would be no stimulation or sense of gain. Ironically, although it appeared that he took his battles less seriously, he actually thrived, living strong and healthy for all his years.

These mythical characters embody two different ways we typically handle change, competitiveness, and life itself. Sometimes we may respond to the environment's seemingly unending demands for change by adopting a "survival" mode—a defensive posture that ultimately exhausts us and those around us. We hang on for dear life trying to keep things the same, avoid change, and get a chance to rest. This chapter can help you enhance your CREATIVE climate for innovation by focusing on "Transition Management."

Even during times of onslaught, we can promote strategically appropriate creativity in a way that *does* rejuvenate us, that gives vibrancy and health to our organizations. This requires that we replace the sometimes unconscious motto of "Protect what we have" with the inspiration to "Create what we want."

Polaroid's founder, Edwin Land, once wrote in an annual report that the company had developed "a Polaroid culture (that gave) an ultimate sense of confidence, purpose, and permanence." Yet with changes in consumer tastes, there came a "mid life" crisis for the company and its culture.[2] One former executive said, "That 'we-can-do-anything' attitude is gone, and I don't think they can recapture it . . . It's a lot more fun to create a market than it is to fight for market share." Polaroid's business strategy later focused on rejuvenating its amateur photography business, broadening its technology base, and diversifying beyond instant photography.

To rejuvenate a business we often must rejuvenate the organizational culture, and rejuvenating the culture can be an important reason for choosing one particular strategy over another! For example, a high-tech firm in the control and instrumentation industry decided to acquire a small robotics company partly because the acquisition might help infuse a new entrepreneurial spirit into the parent company's employees.

Rejuvenation and renewal require a commitment to an organization of excellence, which means a shared commitment to *the highest in both organizational performance and human satisfaction.* Commitment to only one or the other will not do.

The commitment to highest performance and satisfaction is achievable when you exercise the *will* to stretch beyond perceived external and internal limits. It is achievable when you *know* that you can choose and create the circumstances of your work. It is achievable when you are *comfortable with change* as a natural part of life, which author Alan Watts has described as the "wisdom of insecurity." This will, knowledge, and comfort are the source of your power to transform.

This is not as improbable or impractical as you might be thinking.

Transformation and Resistance

The dictionary defines *change* as "to cause to be different; alter; transform; to exchange or replace by another."[3] All organizations go through many changes—reorganizations, automation, personnel turnover, and on and on. Change often brings paradoxical feelings of excitement, fear, longing, or even anger, especially when change is introduced at work. The way the change process is managed often makes the transition more painful than necessary, prompting the perceptions that "We're falling apart!" or "Management doesn't know how to manage."

Organizational growth follows a breathing pattern of expansion and consolidation that corresponds to exhaling and inhaling. When an organization has successfully stabilized a new business with centralized decision making and functional roles (inhaling), the time eventually comes for more decentralization (exhaling). And when decentralization has succeeded in establishing a broader growth pattern for the company (exhaling), there is often a need for more consolidation (inhaling).

The very structure and culture needed to make one stage succeed can become a barrier to later growth. Not realizing this, we often hold back when the limits of one stage have been reached and the next stage is calling for a change in management practices. A typical response is, "Since our old way of managing has worked so far, we just need to do it better." This amounts to the organization "holding its breath." On the other hand, we can also exhale for too long. Change for the sake of change, or change in continual reaction to the environment, eventually diminishes productive energy at all levels. We need a point of stability during change, an anchor of constancy. This is best supplied by an organization's purpose and our commitment to fulfilling that purpose.

For us to hold our "organizational breath" and ignore the need for change means eventual unconsciousness and bare survival at best. To anticipate and embrace necessary change is to breathe long and deeply, thriving instead of merely surviving.

Indeed, there is the danger of trying to minimize the task of organizational change with quick-fix solutions and overly simple programs. The transformation of our work groups, organizations, and larger units of society requires a *critical mass* of support, though not necessarily a majority. This transformation can emerge quite suddenly after the temperature of change has worked up (sometimes subtly) to the boiling point. It might also emerge quite slowly, an evolution that will not be hurried.

There is a dark side to change: the lack of readiness in the organization to change. People in organizations may resist change in any number of forms:

- Being unwilling to take risks

- Dwelling on internal competition

- Remaining focused on short-term goals and operations

- Assigning creativity to a single group, like R&D

- Turning rules and policies into untouchable commandments

- Emphasizing controls even where uncertainties are dominant ("Guarantee to me that this will work.")

Managers and consultants often strategize about how to overcome resistance to change, how to bring about a shift in critical mass. This needs to be looked at very carefully, for there are many different reasons behind resistance, including:

- *Integrity.* You might resist a proposed change out of a sincere belief that it is not the highest good for the people or the problem. This resistance is strengthened when the status quo would be hard to reestablish if the change effort failed. You press for a solution that your heart knows is better.

- *Fear.* You might resist change when you perceive it might threaten your job, status, dignity, budgets, or relationships. You don't trust that the change is really in your best interest. Sometimes this fear may be justified. At other times it may be a "victim" mentality that doesn't like others "controlling" your life. Anger and frustration may spring from the base emotion of fear.

- *Communication.* A lack of understanding of the *need* for change can prevent any serious cooperation, particularly when the price seems too high. A too forceful style of telling people what to do can offend, leading to the same result.

- *History.* You might resist change if you have experienced many meaningless and poorly implemented changes or if you lack confidence in the abilities of the sponsors and change agents to make it work. This resistance is compounded by poor communications.

- *Pace.* You might resist change when you don't have enough time to grieve the passing of the way things used to be. The experience of loss and letting go heals at its own pace. Winters are a necessary time and cannot be hurried.

- *Stress.* You might resist change when there already is too much stimulation for you to manage. You'd rather keep the problems you're familiar with than undergo possible turmoil to get to a "promised land."

> Q What are some examples of times when you resisted change?
> Was your resistance based on any of the above?

The difficulties in transitions come not so much from the extent of change, but from the way change is handled! Each type of resistance is quite natural to our human experience. We do not need either to give in to these resistances passively or to use aggressive, violent force to overcome them. When we honor the reasons for resistance by responding with compassion, we can find active means to melt their snows and create streams of commitment. These means can include education and communication, participation, facilitation and support, or negotiation. The use of manipulation and coercion can back-fire and provoke more stringent resistance. Resistance evaporates when people embrace rather than just understand what would simultaneously benefit themselves and others. Because we don't always automatically know what the greatest benefit for others is, this must be co-explored. This co-exploration is the action that demonstrates proper respect for resistance.

Resistance, therefore, is not something for us to overcome but rather to befriend. (The difference is illustrated by how newspapers reported the first climb to the top of Mt. Everest. A U.S. paper's headline read, "Mount Everest Conquered!" whereas a Japanese paper's headline was, "Mount Everest Befriended.") Often we have to dig deep inside ourselves to forgive ourselves and others for this resistance and promote a movement from fear to forgiveness to forging a future.

The Game of Parts and Wholes

Throughout this book I have proposed that creativity involves recognizing the way things are, including the things we consider dark or negative, and then leaping forward to envision what we want to create. Within our organizations we can best change or transform situations by developing desirable new situations rather than merely fighting undesirable ones. Then, if we find barriers to a more productive and satisfying organization, we can prepare, evoke our inner strength, and create a future instead of burying the present or past.

The need for change often comes when top management develops a new strategy for operating in the marketplace. The strategy is worthless unless your organization can do what's necessary to implement it: producing innovations in products, services, delivery, and operating practices.

Since the parts of our organizations are integral to the whole, we can transform the whole by affecting a critical mass of parts. This is the game of parts and wholes. Within our organizations it is the arena of innovation where we might make our most significant creative contribution.

We need to understand fully the nature of organizational change and then carefully strategize the transition process. This requires more than just a memo or presentation advising people what they're expected to do differently, as the following example illustrates.

For a nationwide insurance carrier the marketplace was dictating the need for a new, computerized claims-processing system. They developed one that could adjudicate 80 to 90 percent of the claims automatically, rather than 40 to 50 percent with the older system. The company found its strategy failing because of poor implementation. In addition to the expected system "bugs," the transition was complicated by the fact that the claims-processing jobs were so different. Performance in previous processing systems was not necessarily a good qualification for the new jobs.

Ultimately, most employees needed to attend a formal training program—at least five full weeks—to work on the new system. Special tests were formulated to determine who would likely succeed in the new jobs. (These tests needed to be validated first.) Employees became agitated because of the validated pretest and the company's use of an employment contract. If employees didn't pass the formal classroom training, they might not be placed in their old job—or even *any* job (because the old system was to be phased out). Employees complained that five to fifteen years of loyal service and good productivity weren't being acknowledged. The "you bet your job" contract also introduced an extraordinary amount of pressure into the classroom learning environment and onto the final exam.

Turnover during and after the formal training periods was twice the normal rate. Consequently, the entire employee-development process was investigated, including recruitment, pretesting, selection, formal classroom training, posttesting, on-the-job training, and on-the-job support. The feasibility of implementing the computer system was also being investigated separately, as were the work-measurement and quality-assurance programs.

Defining the problem correctly was the first major task. Defining it as strictly a morale problem or a turnover problem or a training problem or a computer systems problem or a quality assurance problem would have led to solutions with minimal overall effect. Each group that interacted in the implementation—the supervisors, the trainers, the recruiters, the computer systems people—were making decisions that made all the sense in the world given the pressures they were under. The result of all their actions was a "catch-22" reaction: the more they did to correct the problems, the more something else in the system counteracted the potential benefits.

For example, instead of fully helping to meet production pressures, the transition/employee-development process limited the potential trainee performance. As shown by the following systems flowchart, the need to increase capacity led to decisions for more trainees and more classes per time period. (These were not the only decisions that could have been made, simply the ones that *were* made.) As a result, trainers had few days between six-week courses to prepare and update materials. Given their lack of floor experience and the constant updates on the new systems, their class examples were not always up-to-date. On-the-job training, therefore, had a higher burden for bringing the trainees

Instead of helping to meet production pressures, the current migration and employee-development process on the new computer system limits potential trainee performance.[4]

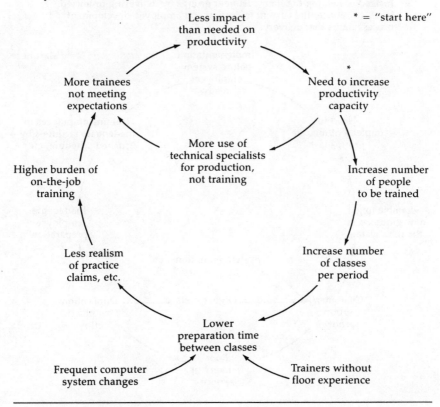

* = "start here"

Less impact than needed on productivity

More trainees not meeting expectations

More use of technical specialists for production, not training

Need to increase productivity capacity

Higher burden of on-the-job training

Increase number of people to be trained

Less realism of practice claims, etc.

Increase number of classes per period

Lower preparation time between classes

Frequent computer system changes

Trainers without floor experience

up to full production. But since the on-the-job trainers (the technical specialists) were needed to help with the production backlogs, they had less time to give. Therefore, more trainees ended up not performing at the levels needed to have the desired impact on productivity.

The "solution" to not having enough trained employees, therefore, led to compounding the problem rather than diminishing it. The problem lay not in any one group not doing its job, but in the way the system of "catch-22" interactions was reinforced by previous decisions and situations.

As another example, instead of helping to retain customer groups by delivering promised levels of service, the transition/employee-development process limited that delivery. With inadequate facilities and employee training, tasks were being done more than once (inefficiencies or correcting mistakes) using the new computer system. Add the processing problems from the new system, and the result was higher costs and poorer service. Customer groups were canceling their enrollments. The marketing staff was scrambling to enroll new groups based on the stated

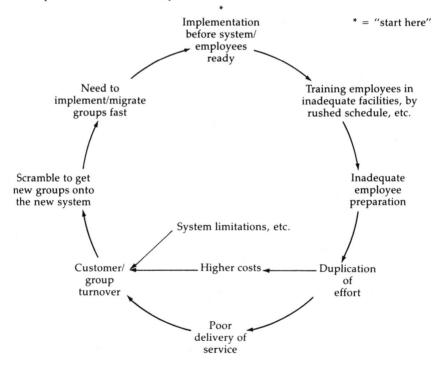

Instead of helping to retain customer groups by delivering promised levels of service, the current migration and employee-development process limits that delivery.

* = "start here"

Implementation before system/employees ready

Need to implement/migrate groups fast

Training employees in inadequate facilities, by rushed schedule, etc.

Scramble to get new groups onto the new system

Inadequate employee preparation

System limitations, etc.

Customer/group turnover ← Higher costs ← Duplication of effort

Poor delivery of service

merits of the new computer system. The need to get these and other groups on the new system compounded the problems based on implementing the system before either the system or the employees were fully prepared.

Again, *each* separate group involved in this process was behaving in very understandable ways given the pressures they were under. The "solution" to getting groups onto the new system quickly compounded the problem rather than diminished it. The problem lay not in any one group not doing its job but in the "catch-22" system of interactions.

By exploring these and three other "catch-22s," the following analogy for possible change strategies emerged.

STRATEGIC TARGETS

Imagine a fast flowing river with two landing ports and jungle in between:

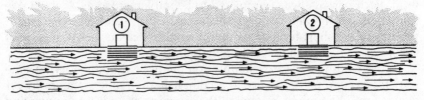

Port #1 is:

"Enhance, build, improve quality"

"Build on positive employee motivations"

"Develop high quality of trainees"

Port #2 is:

Measures, assure, enforce quality"

"Direct policies at hard-to-motivate employees"

"Develop large quantity of trainees"

IF YOU TARGET PORT #2, IT IS VERY DIFFICULT TO GET BACK TO PORT #1.

If we aim for port 2 and land there, it is very difficult to get back to port 1! That means if we develop quality-assurance programs, for example, for the primary purpose of measuring productivity and supporting disciplinary actions, we may have a very difficult time using that program to *improve* quality. In other words, measuring performance is a necessary part of a program for improving quality, but improving quality is not necessarily a part of a program to measure performance. The insurance company was a case of aiming at port 2 *by actions* even though their stated aim was port 1.

The solution to the "catch-22s" was not in changing any one factor, not even in a series of single changes. An organization is like a thermostat: it is set to regulate change around a particular "temperature" of operations. The system is set to respond to any raising or lowering of the temperature by bringing things back to a norm. Only by changing the thermostat setting itself—building a critical mass—can substantive change be sustained.

What was needed at the insurance company to transform the creative climate for adopting the innovative computer system was a package of changes. This package needed to involve *leverage points* that seemed to impact more than one of the "catch-22" cycles at the same time.

Without going into further background detail, the leverage points in the insurance company's transformation effort were: the use of pretesting and classroom tests; the collaborative roles of technical training and quality-assurance personnel; and realignment of classroom/on-the-job training cycles.

The final set of recommendations emphasized a landing at port 1. Eighty proposed actions were grouped into four packages: mandatory short-term, mandatory long-term, optional short-term, optional long-term. Each package was formulated to be implemented as a whole. We cautioned that implementing a few actions from each package might produce some short-term gains, but it wouldn't create a critical mass of substantive change; they would remain in their "catch-22" cycles.

After reading this case example you might be thinking, "What's the use? It's beyond my influence to accomplish any real change in my organization!" You might be feeling that even more so if you're wanting to change a cultural norm (for example, to make your organization "more market-driven"). The paradox you face is that although organizational change can be formidable, it also *is* within your influence when you truly know your own power, love, and wisdom and when you can see your organization as a whole system!

Strategic Innovation Management

In the preceding example we systematically analyzed a particular situation within an insurance company. If we analyze an entire organization and its CREATIVE climate for innovation just as systematically, we notice that there are opportunities for us to exert our influence and promote change.

It is obvious that organizations must relate to a large and complex social, economic, political and technical environment. Within our work situations, we must relate to an equally complex environment.

The first key question we must ask is:

1. What is required of me to relate to, contribute to and be sustained by the environment I operate in?

And as we look around at the people we work with, the next major questions for each of us are:

2. What do I know and expect of myself and the individuals around me?

3. What purpose and goal(s) bring us to work together?

Next, our attention naturally turns to:

4. How do we collaborate and communicate with each other?

Finally, we must face:

5. How do we plan for and prepare to achieve our purpose and goal(s)?

6. How do we implement these plans?

This brings us back to the environment: Have our actions for implementing these plans helped us to relate well to the environment?

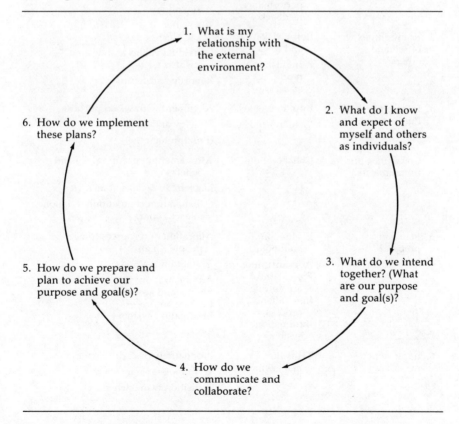

These six questions describe a cycle that provides a framework for understanding how we personally experience work within our particular organizations. This cycle is graphically illustrated on page 209, and is also outlined below. These same questions can also describe our relationship to life in general. Each of these six factors is also directly influenced by a particular element of the CREATIVE climate for innovation. Their relationship is depicted on page 211. However, the model is not complete. There is yet another set of organizational factors that contribute to the climate for innovation:

Personal Experience in an Organization	Related CREATIVE Issues	Detailed Organizational Factors
1. Relate to external environment?	Environmental monitoring Administration: performance feedback	Key Success Factors (for an industry) Environmental monitoring Performance feedback (groups & individuals)
2. Know and expect of self and others?	Roles, risks, and rewards Administration: reporting structure	Talent base Individual values and desired rewards Reporting structure
3. Intend together? Purpose and goals?	Intuition/logic Vision/Purpose	Alignment with vision/purpose Strategy and goals Intuition/logic
4. Collaborate and communicate?	Collaboration	Attunement of group values and beliefs Leadership & delegation styles Collaboration/competition (relation to each other)
5. Plan and prepare?	Evaluation methods Administration: Budget and accounting Innovation process Information systems	Allocation of resources (time, people, finances) Evaluation methods Administration: Budget and accounting Innovation process Information systems
6. Implement plans?	Transition management	Productivity of each function Technology management Transition management

STRATEGIC INNOVATION MANAGEMENT

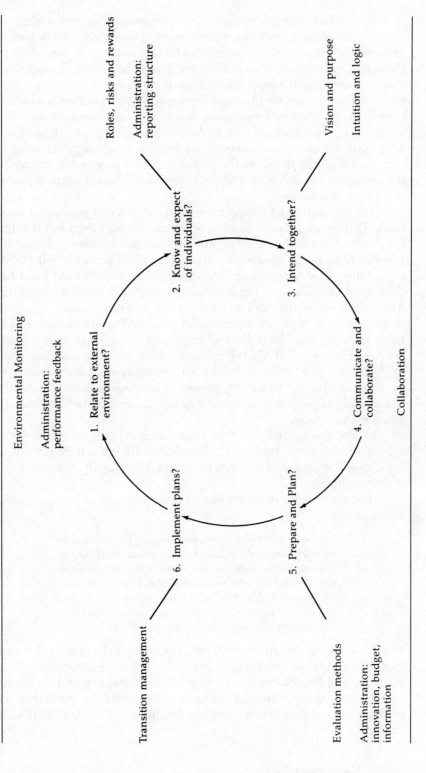

Roles, risks and rewards

Administration:
reporting structure

Vision and purpose

Intuition and logic

Environmental Monitoring

Administration:
performance feedback

2. Know and expect
of individuals?

3. Intend together?

1. Relate to external
environment?

4. Communicate and
collaborate?

Collaboration

6. Implement plans?

5. Prepare and Plan?

Transition management

Evaluation methods

Administration:
innovation, budget,
information

Transitions and changes within our organizations require a shift in a critical mass of these eighteen self-reinforcing organizational factors. Together they define a more detailed model for strategically managing and changing an organization's climate for innovation. The Strategic Innovation Management Model is found on page 213.

With this model for Strategic Innovation Management, you see the complexity of effectively managing the climate for innovation and of developing a critical mass of change for transforming the existing system. This systemization can, however, simplify your strategy for *what* and *how* to change. By using this model as a checklist, you can ensure that the impact of change in one factor is reinforced by supporting changes in other factors as well.

One important fact to remember: not every factor is of equal importance. During an early assessment of organizational needs and readiness for change, certain leverage points will emerge—"make-or-break" factors that must be handled well. An organizational survey called the Strategic Innovation Management Assessment Profile (SI-MAP) has been developed to analyze an organization's innovation climate according to these eighteen factors and to identify the leverage points. Using two databases (one from an organization's SI-MAP respondents and the other from the combined SI-MAP responses of all organizations within the same industry) SI-MAP determines how an organization's profile might be improved through reinforced leverage point transitions.[6] There are three factors that frequently appear as leverage points: alignment on vision and purpose; attunement to group values and beliefs; and transition management.

As discussed earlier in this book, alignment means that people agree on what their purpose is and specifically what they want to become over the course of a number of years. Without alignment we work at cross-purposes.

Thomas Watson, Jr. has made an interesting observation regarding attunement:[7]

> *The basic philosophy, spirit, and drive of an organization have far more to do with its relative achievements than do technological or economic resources, organizational structure, innovation, and timing. All these things weigh heavily in success, but they are, I think, transcended by how strongly the people in the organization believe in its basic precepts and how faithfully they carry them out.*

Finally, to rejuvenate an organization, the ultimate leverage point might well be employing this type of strategic thinking in managing transitions. Uncoordinated change in a few of the eighteen factors is likely to produce temporary change(s) within an organization—a short-term temperature rise—rather than a complete readjustment of the climate's thermostat.

THE STRATEGIC INNOVATION MANAGEMENT MODEL

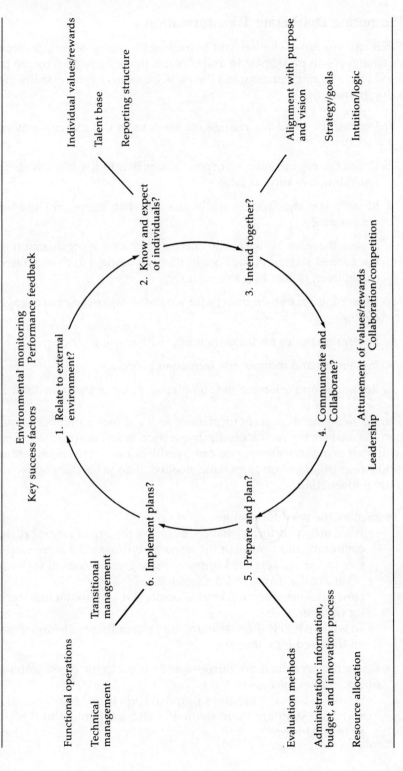

Promoting Deliberate Transformation

What can you do—whether you're a manager, professional, secretary, or whatever—to participate in and constructively assist the change process? First you must understand the eight elements of successfully managing transitions:

1. Explore the need for change (in relation to the external environment).

2. Relate the organization's purpose and vision to the *business strategy* (including key success factors).

3. Identify the organization qualities needed to implement the business strategy.

4. Assess the gaps between the current state of the organization and the needed state, and the organization's readiness to change (across all eighteen factors listed previously).

5. Determine a *transition strategy* for whatever organizational change is needed.

6. Determine tactics for implementing the transition strategy.

7. Implement and monitor the transition process.

8. Recognize and celebrate the completion point of the transition.

The following list may seem formidable as it outlines a number of questions to answer for each step. Each question is important. Particularly with your own work group, you can use the questions to encourage and guide your organization in making needed change less disruptive and more rejuvenating.

- Explore the need for change
  - What are the primary challenges facing the organization? (Insure continuity and control of the ongoing business? Be more responsive to the marketplace? Improve management control systems?)
  - What are the driving forces for change?
  - How does management tend to respond to environmental signals for change?
  - To what extent is there a shared understanding of and agreement on the need for change?

- Relate the organization's purpose and vision to the *business strategy* (including key success factors)
  - What are the organization's purpose and vision?
  - To what extent are there alignment and attunement to this purpose and vision?

- What are the *long-term* purposes/goals of the organization (a qualitative description, more than just quantitative goals)?
- What are the prioritized strategies for responding to the challenges and opportunities of the external environment?

- Identify the organization qualities needed to implement the business strategy.
  - What is the overall vision of the new state of the organization needed to implement the business strategy (explore all eighteen SIM factors)?
  - What goals give the rationale and criteria for designing this new "end state"?
    - Strategic goals ("To be able to produce a greater variety of market-driven products and services")
    - Management goals ("To be able to determine productivity and profit more accurately per product or service")
    - Human resource goals ("To be able to provide job opportunities that make better use of individual talents and aspirations")

- Assess the gaps between the current state of the organization and the needed state as well as the organization's readiness to change (across all eighteen SIM factors listed previously)
  - To what extent are the transition process roles designated for "sponsors," "change agents," and "change targets"?
  - What are the key target areas for change in the organization?
  - How disruptive are the anticipated changes likely to be to functions, systems, and people?
  - What is the sponsor's level of commitment to seeing the change process through to the end? To a positive climate for the transition?
  - How are managers responding to the signals for change (accepting, selectively perceiving the signals, distorting or denying data, ignoring or substituting goals)?
  - To what extent do managers and others understand, believe in, and feel committed to this vision of the future state? (They don't have to like it to be committed to making it work.)
  - What are the current culture and capabilities of staff and systems?
  - What resources are available to devote to the transition?

- Determine a transition strategy for whatever organizational change is needed.
  - What are the optimum sequence and pace of change, addressing all eighteen SIM factors?
  - Who is responsible for managing the transition process, and who is responsible for managing the business operations during the transition stage—to whom do they report and what is their authority?

- Who else will be on the team to manage and monitor the transition process?
- What are the criteria and means of monitoring the transition's progress?

- Determine tactics for implementing the transition strategy.
  - Where is there ability and willingness to make a transition?
  - Where is there inability or resistance to making a transition?
  - Whose backing is critical for gaining widespread support for the change?
  - What type of influence and leadership style will be most appropriate for gaining their commitment?
  - What will be the tactics for responding to resistance to the change (education, participation, facilitation, negotiation, co-opting, coercion)?
  - What management mechanisms—budget systems, performance standards, etc.—can foster and reinforce the transition process?
  - How can the following skills for managing change be recognized and promoted in management?
    - Building commitment
    - Managing resistance
    - Using power and influence
    - Developing synergy

- Implement and monitor the transition process.
  - How are the transition strategy and tactics presented to and implemented with the people in the organization?
  - How well does the transition monitoring process identify: progress according to plan; information vacuums; differences in perception (about the transition); and obstacles to the transition? (How are problems resolved?)
  - How synergistic is the transition management team? How flexible are they in implementing leadership styles? How well do they provide resources and express commitment to the transition process?

- Recognize and celebrate the completion point of the transition.
  - How is the "completion point" of the transition process recognized and announced?
  - Did the transition emerge as planned?
    - Did the master transition strategy work effectively for implementing the business strategy?
    - Did the outcome resolve the organizational needs and satisfy employee expectations for the transition?
  - How is the successful transition celebrated or the unsuccessful one improved and healed?

- How is "maintaining and improving upon change" made a part of management's ongoing responsibility?

Often overlooked in the transition process is the final celebration. Many managers acknowledge business success—and implicitly the organization that produced it—but not organizational success in itself. Change can be difficult, and conscious celebration is *critical* to balance the pressures of transformation and renew the cycles of growth. You can upgrade traditional celebrations, such as banquets and trips, to "festivals of the spirit," introducing rituals that bond your work group, energize its members, and reinforce key values.

Conscious celebrations have five key characteristics:[8]

1. Specific values (authenticity, humor, play, empowerment, elegance, spontaneity, and creativity)

2. Symbols

3. Ritual (fresh rather than stifling)

4. Storytelling

5. Special role of leadership in the events' design and implementation

Celebration can be the ultimate key to transforming our organizations. For example, to help turn around two failing divisions of a major housing company, a vice-president sponsored celebration as part of the company philosophy.[9]

> WE VALUE CELEBRATION AND BELIEVE THAT IT
> ENHANCES WORK:
>
> *We publicly recognize achievements.*
> *We encourage spontaneous celebrations.*
> *We know that if we're having fun, we will work harder,*
> *   smarter, and longer.*
> *We evaluate managers on their enthusiasm and ability to*
> *   create an environment in which it is a pleasure to work.*
> *Employees who spread gloom will be asked to leave.*

It worked so well to inspire work teams that management had to issue a policy requiring employees to take at least one day off per week!

We can make ritual and symbols important in everyday events as well as major transitions. Ron Green, an internal consultant at Alcoa Aluminum, creates celebrations at the end of corporate training programs:[10]

> *During (graduation) dinners, which are light, playful,*
> *fun-filled occasions, we present symbols to each participant*
> *to help celebrate the most important awareness of skill that*

he or she gained during the program—giant ears for listening skills, a clown's nose for appreciating the less serious side of themselves. . . .

These rituals . . . serve to bond the group together, become part of the corporate culture, and act as a reminder that each member has begun a journey toward greater effectiveness.

Conscious celebration depends on the success of the other transition management steps as well. The transition process can be enormously enriching rather than wrenching if managed well. At any level within our organizations, transitions proceed most satisfactorily when each of us who is affected understands and participates in the process. Hopefully, the questions outlined above give you a better feel for what your participation can be.

Where Do I Start?

After all these pages and numerous thoughts, you might be wondering, "Where do I begin if I want to make some changes where I work?" The most pertinent guideline I know of is to *act locally and think globally* about your organization.

Change begins with each one of us planning, intuiting, and acting from whatever position we hold in the organization. Small units hold the organization together, and *you* can make a difference in your work community.

Recall the four faces of the pyramid that symbolizes the key ingredients to your creative edge at work:

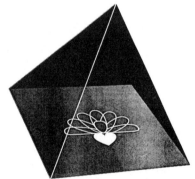

1. The CREATIVE climate for innovation

2. The SPIRITED persons you work with

3. The APPEARE process of creativity·

4. And *you*

Building on the base of your own internal source of creativity, plus your heartfelt values, you can make a difference.

Perhaps the place for you to begin is in developing your own SPIRITED capabilities to be creative. Perhaps it is to sharpen your own APPEARE process for creatively handling situations needing material, idea, spontaneous, event, organizing, relationship, or inner solutions.

Perhaps your next step is to take one aspect of the CREATIVE climate for innovation and focus on developing it around you. You might promote more collaboration, or environmental monitoring, or intuition and logic. Meetings and project work are two areas where substantive improvements in the CREATIVE climate can always be made.

Or perhaps you can take on the transformation of the CREATIVE climate throughout your organization. You might start with an audit of the innovative climate covering each CREATIVE aspect and strategize a long-term change process. This was the approach recently taken by 3M, in which they poked fun at their own reputation for fostering innovation to see if changes were needed.

They found that innovation practices such as the "bootleg slack"—giving employees up to 15 percent of their time to pursue their own projects—was not a reality for workers outside the laboratory. Marketers and manufacturers also had less latitude for mistakes than lab workers. Inadequate rewards for high personal risk, and bureaucratic formalities, also became targets for improvements.[11]

Often the most productive approach to revitalizing an organization is not to focus on the organization but to fix upon a strategy, product, or operation-development project. With a business-oriented task that everyone is invested in accomplishing, organizational transformation takes on new meaning: "How can this organization be improved to implement best the strategy/product/operation plan?" To work together on both business development and organizational development can be your most powerful course of action to achieve impressive bottom-line, top-spirited performance.

Guiding the transformation process takes skill, especially in the key roles of sponsor and change agent. Often, professional consultants, both internal and external to the organization, can play a key change-agent role in promoting and assisting a planned transition process.

IN CLOSING . . .

■ *The important thing is to get people involved, to participate and contribute. I was chairman of a church committee on a project to expand the building, and there was a lot of intense emotion about doing it. Everybody had their own image of what it should be. We held many meetings and kept publicizing it. There might be 1 or 2 people there, or 50, or even 150—but the people bought into the process and then it became theirs. I also learned that it took acknowledgment that issues might be decided for or against that allowed people to continue to contribute.*

The night to vote came after 2½ years. When the meeting moderator called for the vote, the first person called upon was this lady who had invested thirty months of her life into moving this project. She stood up and said, "This place is so special to me, I think it is the most important thing in my life, and that's enough; and I don't want to change a thing. I vote no." When they went around this table, and the vote was obviously going down, everybody was in tears.

The moderator said, "Let's see if we can identify what the problem is." So he split the issues into five pieces to look into each one. Three of the issues passed 13–0 and two of them passed 11–2. So they recalled for the vote on the whole issue, and it passed 11–2 . . . about two hours after the first vote."

Tom, political strategist and former executive for a Fortune 100 conglomerate ◀

Should you protect what you have or create what you want? It's a choice of survival or "thrival." The survival orientation may seem a reasonable, and even necessary, response to legitimate fears. If you aim for protection, you may lose the opportunity to create. But if you aim to create, you can also protect what you value. You can thrive rather than just survive, mobilizing yourself and others with *inspiration* rather than fear.

You can help develop an organization that inspires and empowers people to exercise their talents and values. It depends on how you use your creative APPEARE process, as a SPIRITED person, to promote a CREATIVE climate for innovation and change.

Begin with a firm understanding of how the organization functions as a system, using the SIM model. Develop skills in transition management. Base your actions on both intuitive and linear thinking. Then your commitment to excellence can result in an effective work environment that began as a belief.

You can leverage your insights, skills, and motivations by networking—giving and gathering support for your colleagues, your organization, and yourself to become all you can be. Take any route you can, working within the given structures and also as a part of "parallel organizations" (see chapter 8).

This is not a simplistic formula; the process can be slow, formidable, and complex. Nurture the attitude and belief that you *can* turn your dreams into realistic goals. Master the knowledge and skills it takes to turn realistic goals into reality. Then will you feel the personal power it takes to make a difference.

The choice is yours. You can create an environment, at least with your network of peers and friends, that you *can* invest yourself in. You can live *your* purpose and vision wherever you are—at work, home, anywhere. You can make your purpose and vision the dominant force in the world of your work, rather than waiting for someone else to change things for you.

You have available to you the power, wisdom, love, and integrity to do it. It's inside of you and all around you, perhaps hidden in your workmate or secretary or manager. You can use your creativity to find new ways to tap into that "higher/deeper place" and let the inner vision become the dominant force around you. Your work can be more purposeful, more visionary, and more innovative. It begins with you.

PART IV

Of Profits and Prophets

"*Just how fresh are these insights?*"

Democracy is the idea that every individual is sacred, is a Son or Daughter of God and has inalienable rights to express that unlimited creativity. Science is the mind's understanding of the laws and processes of creation. Technology, when appropriately applied, is the capacity to cooperate with those laws to create a world beyond the limits of sanctity and separation.

—Barbara Marx Hubbard, *A Gift of a Positive Future*

Many of us perceive our organizations as separate and autonomous, engaged in a struggle to survive and thrive. We strategize—even unconsciously—as if we weren't operating in a complex social, economic, and ecological environment. We set goals even when the factors that determine the achievement of those goals are beyond our control and then force ourselves to work under great stress toward all-too-probable failure.

Our organizations are living organisms that join as parts of a greater whole. The purposes of our organizations are best revealed in their interconnection with each other. Like cells in a biological system, we find a meaning for our individual work in contributing to the whole, and this contribution stimulates the optimal flow of nourishment—revenues and profits—to the larger organism.

In addition, every economic and ecological system involves competition as well as collaboration. This applies to systems within an organization (department to department), within a society (organization to organization), and within the world (country to country). But dominance must necessarily begin to destroy the very ecology within which the dominating would wish to thrive. Certainly competition is a great stimulus for people and organizations to grow, to stretch, and to create. But ironically, collaboration is actually the foundation and context within which competition can contribute to the health of organizations.

In this light the religious and revolutionary entreaty to "love your enemy" takes on a new application. Discovering the purpose within the whole and valuing the collaboration that surrounds competition are two approaches to organizational life that hold a higher wisdom than autonomy and dominance. This wisdom lies behind the "excellence" of the most peak-performing, innovative organizations—the organizations that have concluded, along with consultant Roger Harrison, that "strategic thinking is a search for meaning rather than a search for advantage."

The search for meaning must ultimately go beyond goals of "acquiring" and focus on valued contribution to society. As stated in India by the spiritual leader Sai Baba, "Not until man learns to value mankind will anything else find its proper value." This value, this meaning, will be found amidst the creative, productive expression of our minds, hearts, and souls.

CHAPTER 13

Technology with Heart: Bringing Out the Human Values

A Review of Key Themes

A midwestern bank sponsors a multistate movement to "choose a new future" for the mid-American economy—upscaling the quality and quantity of jobs and reestablishing industrial health—through collaborative business, political, and social strategies.

A major Japanese electronics company explores new opportunities for the year 2000 and beyond to help decide what technologies—new materials, new production processes, and so forth—to invest in *now* in order to be positioned for those future opportunities.

A large American computer company explores the barriers to introducing new image-processing systems into the office and manufacturing environments.

A prominent American health insurance company explores new ways to help select and train employees for a totally new automated claims-processing system.

What do these four situations have in common? They are examples of how we relate to technology—how we *envision, produce, use,* and *manage transitions with* new technologies.

These four actions—envisioning, producing, using, and managing transitions—are the links between technology and human values. Like points on a compass, they provide four directions for exploring the relationship between technology and human values.

In this chapter we review many key themes for revitalizing your creative edge, while focusing on developing a closer connection between technology and human values. The principles for developing this intimacy are based on fostering a CREATIVE climate for innovation and change.

As you've seen, creativity develops new ideas that impact and shape our economy, our businesses, and our public institutions. Innovation puts these ideas into action and produces something tangible. It is through creativity and innovation that we envision new futures, develop new products and services, determine what impact we want, and promote transitions.

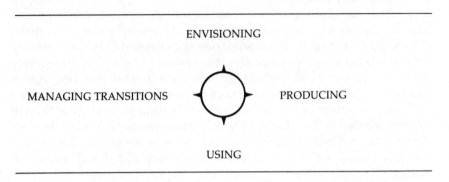

ENVISIONING

MANAGING TRANSITIONS   PRODUCING

USING

The Marriage of Technology and Human Values

Innovative technologies are a critical resource for building a more human future. And the fostering of human values is an economic imperative for the lasting development of innovative technologies.

These are perhaps bold statements—the first in light of the nuclear threat, environmental problems, and dehumanization that people have experienced in our technological world; and the second in light of the seeming polarities between some people in the business community and various subcultures in America.

Technologies *are* a critical resource for a more human future. The fact that diphtheria, cholera, smallpox, and typhoid have been eliminated as prime causes of death is one testimony to this fact. Buckminster Fuller stated in the late 1970s that 50 percent of the world's population at that time maintained a better material standard of living—food, clothing, and shelter—than did 99 percent of the world's population in 1900. As Gandhi, Maslow, and others have pointed out, it is hard to stretch for higher values when essentials for life are missing.

Human values *are* an economic imperative for lasting technological innovation. We must selectively *choose* what we want to develop on the basis of values that build a healthy society and economy and support long-term business and technological development. The primary causes of death in 1983—vascular diseases and cancer—are life-style based; and our life-styles are shaped dramatically by our technologies and the pace of work required to produce them competitively. The flames of increasing health-care costs are being fanned, therefore, by this "technological push" on life-styles. And the health-care burden on our GNP could become fatal to individual industries and our economy as a whole. (For example, health-care costs add over $750 to the price of the average U.S.-made automobile, giving foreign manufacturers a better competitive advantage.) Our economic health will be destroyed if we continue to seek only technological solutions to problems caused to a large degree by a technology-dominated economy. A "value pull" must be added to the "technology push."

What may appear to be technological choices are really human choices. Technology is a mirror of our consciousness and our values. Developments of such "1984" scenarios as computerized psychotherapy or court trials aren't *technical* choices, systems that follow irrevocably from the demands of technology; they are human choices, consequences of people who have lost faith in human interaction and happen to have a computer available. We often still look at technology as though it were inevitable and that it dictates certain solutions. Only when we feel we control technology's development, will we realize that we are not just dealing with technical questions, but social, political, and moral issues.

We have created the world we live in, with its prosperities, poverties, loves, fears, services, and exploitations. We have created how our businesses function, how we compete and collaborate, and how we employ technology.

Technology is a creative art, a function of our tremendous logical and intuitive capabilities. How we envision, develop, and use it is a reflection of our inner wholeness (and dividedness!). Just as "all mind and no heart" produces a barren human being, technology applied without heart produces a barren world and lifestyle—a barrenness that cannot handle the challenges we face in the next decades.

Having "heart" and "human values" can mean a lot of things to different people. Most simply for me, they mean having integrity and caring, expressing our personal and collective talent in ways that benefit and serve others as well as ourselves, and making secular decisions in light of higher spiritual and ethical values.

This is not mere sentiment or a utopian dream but an absolute, practical necessity. We are being pushed toward a new integration of mind, heart, and personal responsibility by many factors,* among them the following:

---

* Personal responsibility is as important as having heart. Heart without a demand for taking responsibility degenerates into saviorism, whereas an emphasis on responsibility, without compassion, degenerates into elitism.

- The globalization of "national" economies

- The competitive environment, including government (de)regulation and inroads of foreign competition

- The pace of technological evolution

- The evolution to an information-based economy

- New and shifting social values/demographic patterns

- Changes in labor force values

- Health, life-style, and stress awareness

- Human survival issues (international relations, hunger, environmental damage, and nuclear destruction)

As we face these challenges with our technology-driven economy, the real question is, "Who's in charge here—people or technology? Does technology serve us, or do we serve technology?" The summary challenge for us is to employ an economics of value(s) that involves our hearts, not just our minds and bank accounts. We *can* conduct our business proactively and elicit the world with our heart's desire rather than "planning" reactively to a runaway future. We *can* steer our organizations to become more effective in meeting these challenges.

Our future will be strongly influenced by the possibilities inherent in our emerging technologies. Today we stand at the crossroads between two very different futures. One is marked by vulnerability, as described above, and the other is filled with opportunity. Technological breakthroughs in areas such as computers, biology, and space can help create new jobs, new opportunities, and new hope. Systems which require a long-term investment in human education and sophisticated equipment will likely be the strength of information-based economies even as the less-developed nations become strong competitors in low technology fields.

Technology and human values *can* become intimate partners (and in many cases already have), for "economics" is most fundamentally the "exchange of value"—or exchange of *values*. Land, products, services, labor, and money are all exchanged in relation to the values of the people involved and the value they place on the *experience* of having those items.

This exchange of value can get seriously out of balance when "paper profits" start to dominate decision making (in takeovers, tax-shelter moves, and such) or when people remain in jobs to "survive" (i.e., have an income) without a fair return for their contributions. With such "trade deficits," employees often impose "import restrictions" on management communications and motivators. Productivity, creativity, and committed involvement of employees can shrink to bare minimums, often focused on ways to *beat* the system rather than enhance it.

Two key questions are:

1. Do we experience the exchange of value between people—individually and in organizations?

2. Are we enough in touch with our environment to know what we give to it and what it gives to us?

Without such feedback, without a tangible experience of value exchange (economy), we live in a fantasy world out of touch with the consequences of our actions.

Numbers are one way of confirming that an organization is producing something that other people value. Volume (of sales and services) is a common measure of the value customers perceive in an organization's products or services. Profits are a measure of an organization's *effectiveness* in producing the value it exchanges. But our inability or unwillingness to value or understand nonquantitative measures of success has kept us thus far from taking charge of our technologies. Rather than making human choices, we let the numbers dictate our actions in envisioning, producing, using, and making transitions with technology.

A Context for Leadership

The partnership of technology and human values requires a specific context for leadership in our organizations. For the moment at least, the window of opportunity is open, and the optimistic future is reachable. But reaching it will require changes in management behavior throughout our institutions. Questions of management controls, motivation, competition, personal power, ethics, and decision making all require redefinition.

For example, motivational management and the design of idiot-proof jobs (which leave little room for creativity) are proving less and less valid in today's work environment. The alienation, isolation, afflictive stress, and even burnout felt by people working under mechanistic management practices are finding expressions in productivity troughs and employee turnover.

In contrast to control-minded *managers, leaders* must exercise power through a shared purpose and vision. Sometimes the vision is first spoken by a key executive and then endorsed by employees. Sometimes workshops are the vehicle for developing that vision.

In the end, however, the vision must emerge from our desire to build a company that harmonizes with our own personal visions. Most often we dearly want to be part of a company that we can invest ourselves in, a company where we can make a heartfelt contribution and be recognized for it. Only then can we rest assured that we are making decisions for the benefit of our whole organization, and that our innovations support an overall sense of direction.

True leadership recognizes that our organizations are not machines whose precise movements can be figured out and manipulated. Rather, organizations are organisms that live by both rational and intuitive means. Management's emphasis on *motivating* individuals is shifting to an emphasis on *empowering* them.

The old work ethic has itself been transformed into a new, more responsible form, emphasizing individual responsibility over order following, personal expression (creativity) over fitting in, and innovativeness over a business-as-usual approach. This new work ethic emerges from our American entrepreneurial traditions, but traditional management practices do not support it.

Social Technologies

The key to revitalizing ourselves, our organizations, and our economy is *choosing a future*. That means more than having a nice dream. It means chasing a future and taking action. For each of the four situations given at the beginning of this chapter—envisioning, producing, using and managing transitions—there are "social technologies" that are useful in enrolling people to build a desired future.

Before discussing these four situations, it is important to recall that there is a spectrum of possible responses to the changing world around us. Sometimes we are *inactive*, ignoring signals of change and failing to act. Sometimes we ignore the signals and only *react* when forced to by crises. At other times we note the signals and move to *plan* actions appropriately. And occasionally we are able to anticipate the signals and move *proactively* to bring about a future we desire. Most organizations respond to signals of change in the inactive and reactive modes; only a few do so in the planning and proactive modes. Yet proactive responses have the best chance of success.

Envisioning

Developing the skill to envision our chosen future is critical, particularly when we are trying to get groups of people to act collaboratively. This is perhaps even more true when dealing with a regional economy or a large, somewhat decentralized organization.

In 1984 SRI International conducted a study commissioned by the AmeriTrust Corporation on how to revitalize mid-America's economy. In the AmeriTrust summary of the study, *Choosing a Future*, the chairman of AmeriTrust, Jerry V. Jarrett, commented:[1]

> *In recent years, [the MidAmerican] economy has not kept pace with that of the nation as a whole . . . MidAmerica can to a large degree take charge of its own destiny . . . We can take the necessary steps to convert cyclical gains into a durable economic renaissance, with more and better jobs for our people. Or we can go ahead with "business as usual" and see each succeeding economic recession further erode our economic base.*

The report also states:

> *There are clear opportunities for the private and public sectors to take major steps that will help restore Mid-American competitiveness and revitalize its key industries. These include: technology utilization, management practices, labor relations, education and training, and capital investment.*
>
> *If decisions are made in the context of the region's economy as a whole . . . the result will be to revitalize the economy and reduce its vulnerability to business cycles. If those decisions are based on what is easiest over the short term . . . MidAmerica will continue to fall behind other areas of the country.*

This statement underscores a critical feature of a healthy, free-market economy. We often assume that competition is the key to a healthy economy, and collaboration must occur within this context. Actually it is the other way around. A vibrant economy on any scope—local, regional, national, international (even "organizational")—depends on competition occurring within the context of collaboration.

This may seem unrealistic in light of industries shutting down or laying off workers due to foreign competition, but that very situation is evidence of what happens when competition becomes the general context for economic relations. Another example is the worldwide recession in the 1970s brought about through free-market control of oil by a few nations. By contrast, working for the "highest good of all" means that collaboration is the context for healthy competition.

Risk and change have a similar relationship. We often fear change because it seems risky. Yet proper change is the key ingredient to future security (not risk) in a changing social and economic environment.

Producing

Many organizations have identified the need to be more innovative. Usually these innovations are intended either to develop and produce new technologies or to make better use of available technologies.

Companies in mature industries face the choice of either changing and innovating or dying. Others are successfully riding high for now but know that they need to be doing something different in the next five and ten years to remain profitable. Toshiba, as we saw earlier, has begun looking twenty to twenty-five years ahead to discover what technologies they might invest in *now* to give them the best chance of being *positioned* properly for future business opportunities.

As many companies begin to plan for their future, their first questions usually are, "What will people buy? What will they want? What will they value?" And then they initiate planning to react or respond to (usually) a single forecast of what the markets will be like. The balancing questions, not so frequently asked, are, "What do we *want* to be? What do we stand for? What are we *called* to be?" We need to evoke our alignment and attunement to guide our decision making and inspire our personal investment of energy.

In order to effectively foster new ideas and carry them out to completion, we need to do more than just learn new brainstorming techniques, introduce better incentive programs, or develop "intrapreneurial" ventures. A climate conducive to developing and acting on new ideas needs to be fostered.

The eight principles discussed throughout this book that contribute to a positive, CREATIVE climate for innovation are:

C = COLLABORATION, COMMUNION, AND COMMUNICATION (across groups)

R = ROLES, RECOGNITION, RISK, AND REWARDS (of individuals)

E = ENVIRONMENTAL MONITORING (and widespread distribution of information)

A = ALLEGRO ADMINISTRATION (of organizational process and systems)

T = TRANSITION MANAGEMENT (for organizational change)

I = INTUITION AND LOGIC (honored in idea generation and decision making)

V = VISION AND ROLE OF INNOVATION (communicated and shared)

E = EVALUATION METHODS (for targeted innovation)

There are many approaches to implementing each of these principles in an organization. The art and science of such work involves an understanding of where the business is heading and how the current organization can or cannot support that vision.

Using Technology

Productivity is based on using our talents to provide a mutually satisfactory exchange of value. Innovation and human values are what keep the exchange of value alive and well.

When technology is introduced into our work settings (usually, to improve productivity), it can have a dramatic impact on us and our work. A realistic and timely environmental impact report for a new technology is critical to its successful buy-in and implementation.

There are three key questions for such an assessment:[2]

1. Is management prepared to push the system through?

2. How fast can the system be put into full operation?

3. Are the current skills and knowledge of the employees compatible with the needs of the new system?

Interviews are conducted to determine the fit between people's perceptions of their jobs and the proposed technology system. A *fit* is defined as the employees seeing the system benefiting them *personally*. A lack of fit is characterized by expectations of lost jobs, meaningless work, and low management commitment. Under such circumstances, successful implementation is not likely and is considered a high risk.

You can identify areas of low, medium, and high risk relative to people's ability and willingness to implement the system. When appropriate, you can promote training programs or policy changes to make the fit and impact more desirable. You can use this type of management impact assessment with up-and-running systems that are in trouble or with proposed systems as a preventative measure.

Managing Transition

We all can experience anxiety and resistance to the change that is inherent with innovation, or we can embrace change with a sense of personal responsibility to guide it. Resistance to change can stem from employees or managers (decision makers at all levels of the hierarchy) for many reasons. Disruption increases as the traditional assumptions and norms no longer "work" in the new environment.

Throughout the transition process, especially with the implementation, there are some generic skills you can practice to nourish the process:[3]

- Building commitment

- Managing resistance

- Using power and influence

- Developing synergy

- Focusing on sponsors (those who want the change), agents (those who catalyze the change), and targets (those who are most impacted by the change)

Careful planning and facilitation of the change process can humanize the transition, produce a new way of working, and make it much more cost-effective and productive. There are three prime issues in managing transitions: *readiness, planning,* and *implementation.*

An organization's readiness to change is composed of:

a. the relative dissatisfaction and/or vision of the future by the person(s) desiring the change,

b. the potential resistance of those most affected by the change,

c. the commitment of the sponsoring person(s).

In a., dissatisfaction may result because your organization is not functioning well and future success depends on change. Even if your organization *is* functioning well, change may still be necessary for future success. In b., the resistance you might face can be based on values and emotions, demonstrated through political, economic, and logistical struggles. In c., your sponsors must be ready to commit resources, exercise public and private visibility, and persist until the depth of change is achieved.

Transition plans can have both strategic and tactical elements. The strategy includes descriptions of discrepancies between current and desired states; the necessary changes, sequence, and timing of events; and criteria for acknowledging success. The tactics include identifying key people for support and resistance points; assessing levels of commitment needed from key people; determining power and influence styles; designing reinforcement systems; and determining how synergy is best promoted.

How you implement your strategy and tactics *through people* is critical to success. You may wish to assign people specific roles using "tell" or "sell" leadership styles (see chapter 9), or enroll them with "consult" or "participative" styles. While the latter two styles are often touted as the human-values ideal for all cases, there are two factors that may point toward the use of the former. The first factor is *time:* the time it takes to reach a group decision may be longer than a crisis allows. The second factor is *resistance to change:* Realistic concerns about job security, for example, need to be addressed while not giving in to "victim" thinking ("The change is *making me* feel anxious").

IN CLOSING . . .

Often we perceive ourselves as victims of our technologies, our work environment, our economic structure, and our managers. We then strive to get these "causes" of our ennui to change so that we can feel powerful rather than victimized. But even if that works, we still perceive ourselves as victims because we believe the environment causes our feelings.

Several years ago I listened to a nun, who had just returned from a memorial service in Hiroshima, speak about eliminating the nuclear threat to the world. She reminded us not to fall into the trap of "them versus us"—those who want bombs versus those who don't—because that same consciousness gives rise to war in any form. Instead, she suggested we remember that we have our own 'departments of defense' between ourselves and others, that we have our own 'security councils' keeping us safe, that we have our own 'departments of energy' determining how we use our energies. The departments in Washington are but a reflection of our own values and tendencies. She concluded that from a spirit of union and with a sense of personal responsibility we can rise together to accomplish goals of safety, security, and energy management.

Substantive change—transformation or evolution—occurs with less effort when we have an inner sense of power and responsibility for events *first*, then take action that has a chance of making a difference. The power enables our actions to be more effective. A victim's actions, with an underlying, self-perceived powerlessness, usually produce self-fulfilling prophecies of little real change.

For many of us the source of this inner power comes from our spiritual experiences and ethical principles. From that basis in integrity we are more willing to share each other's weaknesses and burdens and climb towards higher values and prosperity (in all meanings of the word) together. We therefore accept more responsibility for contributing to our organizations and society. We realize that as we share each problem together, we achieve together.

Each one of us is creative, wanting at some level to find ways to express who we are and what we stand for. Our organizations are like us, and as groups of 10, 100, or 10,000 we want to know who we are, what we stand for, and how we can express ourselves creatively in the world. We each have a role to play in our organizations to make them, and us, more financially and emotionally prosperous. The personal and organizational challenge before us is to be not only more responsive to opportunities but also more proactive in bringing about those opportunities.

We are by nature creative. Creativity is the expression of our talents, our hearts, our essence, individually and in groups (organizations). At our most efficient, most productive, most peak performing, we envision with heart, we produce with heart, we use with heart, we transition with heart and we endow our technologies with heart. There we live most vitally, most prosperously, most humanly.

CHAPTER 14

You and the Family

The business of America is Business.

Calvin Coolidge

We are created to develop the ability to create. The creature is designed to mature into the creator, the Son into the Father. The Creation is the way by which God the One becomes many, and why Eternity is in love with the productions of Time.

Joseph Chilton Pearce, The Bond of Power

To some, these statements are antithetical or contradictory. Certainly, their focus is different . . . or is it? Business in its basic form is the exchange of value(s) among people. The *value* is represented by the product(s) of our individual and collective creations. Where do these creations come from? How do they come about?

One answer is that we generate them from the inside, from our dreams and ideas, our longings and visions. We produce them and give them to the world, often by our collective alignment and attunement. We apply our personal talents and collaborate in work systems designed to help us achieve those visions. But the full answer to these questions lies deeper.

St. Augustine said in his *Confessions,*

> All the loveliness which passes through men's minds into their skillful hands comes from that supreme loveliness which is above our soul, which my soul sighs for day and night.

And Tolstoy said in *What Is Art?,*

> Art is a human activity, whose purpose is the transmission of the highest and best feelings which men have attained.

The statements of Coolidge and Pearce merge with those of St. Augustine and Tolstoy in the realization that business is a form of art. The consummate artist demands a level of dedication and practice beyond what most people in organizations dream of applying to themselves and their work. Business as art demands this same level of personal mastery. And without personal mastery of whatever God-given talents we each have, we are left with only a quiet voice that wants to be expressed, a "muscle" of creativity that wants to be exercised. Andre Previn once said, "If I miss a day of practice, I know it. If I miss two days, my manager knows it. If I miss three days, my audience knows it."

The artistry of business is the creative right of all those who desire true prosperity—economically, spiritually, personally, and globally. Are you up to it? Will you stand up for it?

We now have a truly global economy; no nation is an island unto itself, unaffected by the economic problems or successes of other nations. Our technology has linked us; communication between persons virtually anywhere on the planet is possible. Our awareness of being part of a global family is, however, just beginning to catch up with these developments. Our human family includes both the person on the other side of your desk and the person on the other side of the world. This awareness will have profound effects on all phases of our life and work.

We *are* the organizations we work for. Companies as diverse as IBM, Frito-Lay, Walt Disney Productions, and the Smith and Hawkins Tool Company in California have built their success with the help of employees who seriously believe that "We are the Customer: we treat them as we would want to be treated. We are the Company, responsible for everything that happens here, and we share in every success. And we make sure that we are receiving at least equal to what we are giving in our work."

And we are not only part of our own company; we're also stakeholders in the actions of each and every organization. If we choose to let it, the future we create together can profit each of us. Only when we make our workplaces more creative, more productive, and more alive will we know how to achieve true prosperity—true material well-being and quality of living.

The greatest challenge facing each of us is mastering our talents, and using those talents for the prosperity of whatever group(s) we identify with—our department, organization, country, and/or world. That is the challenge behind Tom Peters's notion that the bottom line in excellent organizations is "love and passion—loving what you do and doing it with passion."

We typically spend around 2,000 hours per year at work, sometimes more. This represents 30 to 40 percent of our waking lives. All the joys and difficulties of life can occur during these work hours. Because our jobs are a major factor in our lives, our organizations are ideal places to begin making society a place where each person can express his or her best. Author Brian Swimme says,[1]

> *Precisely because you are aware of the limits of life, you are compelled to bring forth what is within you; this is the only time you have to show yourself.*
>
> *You can't waste away in a meaningless job, cramming your life with trivia; The supreme insistence of life is that you enter the adventure of creating yourself.*

And Kahlil Gibran adds:[2]

> *Work is love made visible.*
> *When you work with love,*
> *you bind yourself to yourself,*
> *and to one another, and to God.*

Grow yourself. Create a nurturing environment for yourself. Commit yourself wholeheartedly to your organization's being the best it can be. Don't wait for the organization to prove itself to you; the proof of the worth of your commitment will come *after* the commitment, not before.

Become aware of your organization's external environment and internal climate. Learn new ways to explore problems and entertain new possibilities. Learn more about being a visionary leader, taking the initiative, and building coalitions. Get others involved in your vision, get involved in theirs, and persist until collectively you are successful. Learn how organizational *systems* work to reinforce your values and ease the innovative process. Learn to embrace the changing world—and the world of change—and master the art of managing change.

Develop yourself. Assist others in their development. In the world of life and organisms, if you're not growing, you're dead. Your organization can suffer the same fate. And if you *are* growing, you're celebrating. Life at its fullest is a celebration. And life at work is no exception.

There is much for us to do together to bring about the quality of work life that we know is possible, for the sake of whatever we most value and cherish: better health, higher living standards, a greater sense of family and community, profitable progress, inner richness, or more regard for the whole planet. We can have them all.

> ◧ *With an eye made quiet by the power of harmony and the deep power of joy we see into the heart of things.*
>
> —*William Wordsworth* ◳

See into your own heart, and into the heart of your organization, and into the heart of creativity itself. See for yourself the potential that lies there—a newborn, a prophet, a part of *you*, a part of *your* family.

This book has been dedicated to a world beyond scarcity and separation, to a world where each person prospers materially and spiritually. Ultimately, the value of this book is not in what it says, but in what people like you do with the thoughts and information it contains. What you do and achieve is your gift to us all. Thank you.

An Invitation

If you have comments or questions about what I've written, or if you have stories/experiences about being creative at work, I would appreciate hearing them (perhaps to share with other people). Please write to me at P.O. Box 294, Mill Valley, California 94941. Thank you.

William C. Miller

Suggested Readings

On the following pages you will find a sampling of books and articles you might wish to explore.

—WCM

BOOKS

Adams, J. *Conceptual Blockbusting*. 3d ed. Reading, Mass.: Addison-Wesley, 1986.

Adams, J., ed. *Transforming Work*. Alexandria, Virginia: Miles River Press, 1984.

Bandrowski, J. *Creative Planning Throughout the Organization*. New York: AMACOM, 1986.

Bennis, W. G., K. D. Benne, Robert Chin, and K. E. Corey. *The Planning of Change*, 3d ed. New York: Holt, Rinehart & Winston, 1976.

Blanchard, K., and P. Hersey. *Management of Organizational Behavior: Utilizing Human Resources*. 4th ed. Englewood Cliffs, N.J.: Prentice-Hall, 1982.

Buzan, T. *Use Both Sides of Your Brain*. New York: Dutton, 1983.

Deal, T., and A. Kennedy. *Corporate Cultures*. Reading, Mass: Addison-Wesley, 1982.

deBono, E. *Lateral Thinking Creativity Step by Step*. New York: Harper & Row, 1970.

deBono, E. *Lateral Thinking for Management*. New York: McGraw-Hill, 1971.

deBono, E. *Eureka: An Illustrated History of Inventions from the Wheel to the Computer*. New York: Holt, Rinehart & Winston, 1974.

deBono, E. *Teaching Thinking*. New York: Penguin Books, 1980.

Doyle, M., and D. Straus. *How to Make Meetings Work*. New York: The Berkley Publishing Group, 1976.

Drucker, P. *Innovation and Entrepreneurship*. New York: Harper & Row, 1985.

Garfield, C. *Peak Performers: New Heroes of American Business*. New York: William Morrow, 1986.

Garfield, P. *Creative Dreaming*. New York: Ballantine, 1974.

Gawain, S. *Creative Visualism*. Mill Valley, Calif.: Whatever Publishing, 1978.

Goldberg, P. *The Intuitive Edge: Understanding and Developing Intuition*. Los Angeles, Calif.: Tarcher Publishing, 1983.

Goldberg, P., and C. Hegarty. *How To Manage Your Boss*. Rawson Associates, New York. 1981.

Gordon, W. *Synectics—The Development of Creative Capacity.* New York: Harper & Row, 1961.

Hampden-Truner, C. *Maps of the Mind: Charts and Concepts of the Mind and Its Labyrinths.* New York: Collier Books, 1981.

Hanks, K. and J. Parry, *Wake Up Your Creative Genius.* Los Angeles, Calif.: Kaufmann, 1983.

Harman, W. *Creativity and Intuition in Business.* Menlo Park, Calif.: SRI International, 1985.

Harman, W., and H. Rheingold. *Higher Creativity: Liberating the Unconscious for Breakthrough Insights.* Los Angeles, Calif.: Tarcher Publishing, 1984.

Joy, W. B., M.D. *Joy's Way.* Los Angeles, Calif.: Tarcher Publishing, 1979.

Kanter, R. M. *The Change Masters.* New York: Simon & Schuster, 1983.

Kaufman, R. *Identifying and Solving Problems: A System Approach.* 3d ed. San Diego, Calif.: University Associates, 1982.

Kelley, R. *The Gold Collar Worker.* Reading, Mass.: Addison-Wesley, 1985.

Krishnamurti, J. *Beyond Violence.* New York: Harper & Row, 1973.

Lao Tsu. *Tao Te Ching.* Translated by Gia-Fu Feng and Jane English, New York: Alfred A. Knopf, 1972.

Maslow, A. *The Farther Reaches of Human Nature.* New York: Viking Press, 1974.

Mooney, R. L., and T. A. Razik, eds. *Explorations in Creativity.* New York: Harper & Row, 1967.

Osborn, A. F. *Applied Imagination: Principles and Procedures of Creative Problem-Solving.* New York: Scribner's, 1963.

Parnes, S. J. *The Magic of Your Mind.* Buffalo, N.Y.: Creative Education Foundation and Bearly Limited, 1981.

Pelletier, K. R. *Toward a Science of Consciousness.* New York: Delta Books, 1977.

Peters, T., and R. Waterman, Jr. *In Search of Excellence.* New York: Warner Books, 1982.

Raudsepp, E. *How Creative Are You?* New York: Perigree Books, 1981.

Rogers, C. *Freedom to Learn.* Columbus, Ohio: Merrill, 1979.

Rogers, E. M. *Diffusion of Innovations.* 3d ed. New York: Free Press, 1983.

Rubin, I., W. Fry, and J. Plovnik. *Task-Oriented Team Development.* New York: McGraw-Hill, 1978.

Shoff, J., J. Connella, P. Robin, and G. Sobel. *Imagery—Its Many Dimensions and Applications.* New York: Plenium Press, 1980.

Smith, A. *The Powers of Mind.* New York: Summit Books, 1982.

Swimme, B. *The Universe is a Green Dragon.* Sante Fe, N.M.: Bear & Company, 1985.

SYDA Foundation, *Creativity and the Self.* South Fallsburg, N.Y.: 1983.

Vaughan, F. and R. Von Oech. *How to Unlock Your Mind for Innovation: A Whack on the Side of the Head.* Menlo Park, Calif.: Creative Think, 1982.

Watson, T., Jr. *A Business and Its Beliefs: The Ideas That Helped Build IBM.* New York: McGraw-Hill, 1963.

ARTICLES

Carne, G. C., and M. J. Kirton, "Styles of Creativity: Test Score Correlations Between Kirton Adoption-Innovation Inventory and Myers-Briggs Type Indicator." *Psychology Reports* 50 (1982): p. 31–36.

Douglas, J. H. "The Genius of Everyman." *Science News* (April 23, 1977): p. 268–270.

Gibson, C., and C. Singer, "New Risks for MIS Managers." *Computer World* (April 1982), In Depth, p. 1–11.

Mulligan, G., and W. Martin, "Adaptors, Innovators, and the Kirton-Innovation Inventory." *Psychology Reports* 46 (1980): p. 883–92.

Park, F. "The Technical Strategy of 3M: Start More Little Businesses and More Little Businessmen." *Innovation* 5 (1969); p. 12–26.

Pines, M. "Psychological Hardiness: The Role of Challenge in Health." *Psychology Today* (December 1980): p. 34–44, 98.

Schmidt-Tiedemann, K. J. "A New Model of the Innovation Process." *Research Management* (March 1982): p. 18–21.

Smith, E. "Are You Creative?" *Business Week* (September 30, 1985): 80–84.

Turecamo, D. "Creativity and the Business Manager." *Skylite* (June 1984).

Wack, P. "Scenarios: Uncharted Waters Ahead." *Harvard Business Review* (September-October 1985): p. 72–89.

Wack, P. "Shooting the Rapids." *Harvard Business Review* (November-December 1985): p. 139–150.

Notes

INTRODUCTION

1. William R. Hewlett and David Packard. *The HP Way.* Palo Alto, Calif.: Hewlett-Packard, 1980, p. 3.

2. *The American Heritage Dictionary of the English Language.* New College Ed. Boston: Houghton Mifflin Co., 1981, p. 311.

3. Dorrine Turecamo. "Creativity and the Business Manager." *Skylite*, (June 1984): 46.

PART I *Fostering Creativity and Innovation from Anywhere in the Hierarchy*

CHAPTER 1 *You and Your Creativity*

1. Philip Goldberg and Chris Hegarty. *How to Manage Your Boss.* New York: Rawson Associates, 1981, p. 1–2.

2. See Robert Kelley. *The Gold Collar Worker.* Reading, Mass.: Addison-Wesley, 1985.

3. James Kouzes and Barry Posner. "When Are Leaders at Their Best?" *Santa Clara Magazine* (Winter 1985): 4.

4. Gifford Pinchot coined the term "intrapreneur" to refer to, "Those who take hands-on responsibility for creating innovation of any kind within an organization . . . The dreamer who figures out how to turn an idea into a profitable reality." He defines an entrepreneur as, "someone who fills this role of intrapreneur outside the organization."

5. Dorrine Turecamo, "Creativity and the Business Manager." *Skylite* (June 1984): 46.

6. Kurt Hanks and Jay Parry. *Wake Up Your Creative Genius.* Los Angeles, Calif.: Kaufman, Inc., 1983, p. 21.

CHAPTER 2 *You and Your Organization's Climate for Creativity*

1. Harlan Cleveland. "Information as a Resource." *The Futurist*, (December 1982): 35.

2. From a speech before the American Society of Newspaper Editors, April 16, 1953.

PART II   *Individuals Working Creatively*

CHAPTER 3   *Developing Yourself as a Creative Individual*

1. Willis Harman and Howard Rheingold. *Higher Creativity: Liberating the Unconscious for Breakthrough Insights.* Los Angeles: Tarcher Publishing, 1984, p. 5.

2. Banesh Hoffman. *Albert Einstein: Creator and Rebel.* New York: New American Library, 1972, p. 99.

3. From J. H. Douglas. "The Genius of Everyman." *Science News,* (April 23, 1977): 268.

4. Eric Adams. "Electric Pets." *American Way.* (March 18, 1986): 70–72.

5. W. Turner. "How the IBM Award Program Works." *Research Management* (July, 1979): 24–27.

6. *One Hundred Poems of Kabir,* translated by Rabindranath Tagore. London: Macmillan Publishers Ltd., Copyright © 1961, p. 75.

CHAPTER 4   *Enhancing Your Individual Creative Process*

1. Hoffman. *Albert Einstein,* p. 124–25.

2. From "An Essay on Criticism," *A Treasury of the World's Best Loved Poems,* New York: Avenel Books, 1961, p. 21–22.

CHAPTER 5   *Using Linear Techniques for Idea Generation*

1. Patricia Garfield. *Creative Dreaming.* New York: Ballantine, 1974, p. 44.

2. Hanks and Parry. *Wake Up Your Creative Genius,* p. 55.

3. Thanks to Scott Isaksen, Director of Creative Studies, State University College at Buffalo, for this list.

4. Pierre Wack. "Scenarios: Uncharted Waters Ahead." *Harvard Business Review,* (September-October, 1985): 72–89.
Pierre Wack. "Shooting the Rapids." *Harvard Business Review* (November-December 1985): 139–150.

5. Used with permission of the company.

CHAPTER 6   *Using Intuitive Techniques for Idea Generation*

1. Thanks to Sharon Jeffrey-Lehrer for this example.

2. Thanks to Chuck McConnell for this drawing.

3. Thanks to Neila Miller of People*Systems Potential for this method.

4. Thanks to Jennifer Hammond for this chart.

CHAPTER 7   *Transforming Individual Blocks to Creativity*

1. Thanks to Werner Erhard for his telling of this story.

2. James Adams. *Conceptual Blockbusting.* 3d ed. Reading, Mass.: Addison-Wesley, 1986, p. 4.

3. Ibid., p. 188.

4. Maya Pines. "Psychological Hardiness: The Role of Challenge in Health." *Psychology Today* (December 1980): 34.

5. Adam Smith. *The Powers of Mind*. New York: Summit Books, 1982, p. 188.

6. Lao Tsu. *Tao Te Ching*. Translated by Gia-Fu Feng and Jane English. New York: Alfred A. Knopf, 1972, p. 63.

PART III *Groups Working Creatively*

CHAPTER 8 *Establishing Your Role in Creative Groups*

1. The latter approach is typified by a program from I. Rubin, W. Fry, and J. Plovnik. *Task-Oriented Team Development*. New York: McGraw-Hill, 1978.

2. Thanks to Dennis Jaffe for this formulation.

CHAPTER 9 *Taking a Lead in Group Problem Solving*

1. Michael Doyle and David Straus. *How to Make Meetings Work*. New York: Berkeley Publishing Group, 1976.

2. Dr. Joseph McPherson originally developed the procedure described.

CHAPTER 10 *Business Innovation with a Purpose, Vision, and Strategy*

1. Thomas Watson, Jr. *A Business and Its Beliefs: The Ideas That Helped Build IBM*. New York: McGraw-Hill, 1982, p. 4–6.

2. Thomas Peters and Robert Waterman, Jr. *In Search of Excellence*. New York: Warner Books, 1982, p. 279.

3. Terrence Deal and Allen Kennedy. *Corporate Cultures*. Reading, Mass.: Addison-Wesley, 1982, p. 37, 43.

4. Marie Spadoni. "Eureka! A Lot of Prospecting Is Behind Research Breakthroughs." *Advertising Age* (October 31, 1983): 52–3.

5. James Bandrowski. *Creative Planning Throughout the Organization*. New York: AMACOM, 1986, p. 17.

6. Gendron, George. "Bitter Victories." *Inc.* (August 1985): p. 25–27.

7. Kotkin, Joel. "Why Products Fail." *Inc.* (May 1984): p. 50–5.

CHAPTER 11 *Institutionalizing Innovation Where You Work*

1. K. J. Schmidt-Tiedemann. "A New Model of the Innovation Process." *Research Management*, (March 1982): 18–21.

2. Used with permission of the company.

3. R. Rosenfeld and J. Servo. "Business & Creativity: Making Ideas Connect." *The Futurist* 17 (1984): 21–26.

4. Rosenfeld and Servo. "Business and Creativity."

5. In a letter to the author dated November 13, 1985.

6. Bernie Ward. "Center of Imagination." *Sky* (June 1985): p. 22–80.

7. Rosabeth Moss Kanter, *The Change Masters,* New York: Simon & Schuster, 1983, p. 19, 372–76.

8. Ford Park. "The Technical Strategy of 3M: Start More Little Businesses and More Little Businessmen." *Innovation* (1969): 20.

9. Kanter, *The Change Masters,* p. 28–35.

10. Published by 3M Company, St. Paul, Minn.

11. Kabir. *One Hundred Poems of Kabir.* p. 56.

CHAPTER 12   *Rejuvenating Your Workplace through Planned Change*

1. Thanks to Juanita Brown for coining this term.

2. William Bulkeley. "Losing Its Flash." *The Wall Street Journal* (May 10, 1983): p. 1, 20.

3. American Heritage Dictionary. p. 224.

4. Used with permission of the company.

5. Used with permission of the company.

6. For more information about Strategic Innovation Assessment Profile (SI-MAP) write to William C. Miller at P. O. Box 294, Mill Valley, Calif. 94941.

7. Thomas Watson, Jr., *A Business and Its Beliefs,* p. 4–6.

8. Thanks to Cathy DeForest, Ph.D. for sharing her article, "The Art of Conscious Celebration: A New Concept for Today's Leaders." Copyright 1985. p. 4. Published in *Transforming Leadership: from Vision to Results,* John D. Adams, General Editor. Alexandria, VA: Miles River Press, 1986.

9. DeForest. "The Art of Conscious Celebration," p. 11.

10. Ibid., p. 9.

11. Debra Whitfield, "3M Rips Down Bureaucracy to Spur Creativity." *Los Angeles Times* p. D7.

PART IV   *Of Profits and Prophets*

CHAPTER 13   *Technology with Heart: Bringing Out the Human Values*

1. Published by AmeriTrust Corporation, Cleveland, Ohio, 1984.

2. Cyrus Gibson and Charles Singer. "New Risks for MIS Managers." *ComputerWorld* (April 1982): 56ff.

3. See Daryl Connor. *Managing Organizational Change.* Atlanta: Thompson-Mitchell & Associates, 1983, p. 27–29.

CHAPTER 14   *You and the Family*

1. Brian Swimme. *The Universe is a Green Dragon.* Sante Fe, N.M.: Bear & Company, 1985, p. 117.

2. As quoted in Christopher Hegarty's *How to Manage Your Boss.* New York: Ballantine, 1984, p. iv.

Index

Kennedy, Allan A., *Corporate Cultures* (with Deal), 153–154
Kepner-Tregoe, 81
Key-word index, 71
Kilmartin, Jack, 45
Kirton, J. M., 123
Kodak, 183
  Office of Innovation Network (OIN) at, 45, 179–182
Krough, Les, 187

Labor force values, changes in, 20, 227
Lambdek Fiber Optics, 181
Land, Edwin, 200
Lao Tsu, 115
Leadership, 14–15
  context for, 228–229
  and delegation styles, 126–128
  visionary, 153–154
Lehr, Lewis, 187
Lewin, Kurt, 72
Life-style
  awareness of, 20, 227
  shaped by technology, 226
Linear techniques, 65, 66–81, 83, 111
  combination of intuitive techniques and, 98–100, 167, 170
Linear Technology, 151
Lipetz, Philip, 96
Logic, 26
  link between intuition and, 65, 98, 231
  See also Linear techniques
Love and passion, 62, 236

MacPaint, 189
Maidique, Modesto, 171–172
Management practices, as target for innovative thinking, 125, 126
  See also Allegro admininstration; Evaluation process(es)
Market-functions matrix, 67
Market segmentation schemes, initial choice of, 161
Market/technology matrix, 67
Maslow, Abraham, 37, 41, 108, 225
Material creativity, 7–8, 22, 35
Matrix analysis, 66, 67–68
Maugham, Somerset, 17
MaxThink, 189
McPherson, Joseph, 37, 41
Meditation, 84, 94–98
Meetings, problem-solving, 134–136, 148

and problem-solving process, 136–139
  roles and expectations in, 139–142
  See also Workshops, major problem-solving
Melon, Tom, xviii
Mervyn's, 45
Metcalfe, Bob, 15
Michelangelo, 35, 42
Minghus, Charles, xix
MIT, 52
Monitoring
  environmental, 25, 231
  systems, 160–161
Monolithic Memories, 130
Morphological analysis, 66, 68–70, 89

National Training Laboratories, 129
Need-drivens, 127, 161
Networking, 160, 161
Newton, Sir Isaac, 101
North American Tool and Die Incorporated, xviii
"Not invented here" (NIH) syndrome, 107

O. D. Resources, 156–157
Oil crisis (1973), 76, 79
Olga Company, xviii
Oregon, University of, 194
Organization structures, as target for innovative thinking, 125
Organization creativity, 9–10, 22, 35
Originators, 123, 124
Osborn, Alex, 74
Outer-directeds, 128, 162
Owens-Corning, 183, 196–197
Oxford University, 4

Parallel organization, acting as, 131–132, 220
Patent Office, U.S., 4
P'DAGEDIE process, 136–138, 140, 141
Pearce, Joseph Chilton, *The Bond of Power*, 235, 236
People Express Airlines, 171
Performance appraisal, and human resource practices, 184–187
Person, characterizing creative, 36–41
Personal and cultural perceptions, as blocks to creativity, 108–109
Peters, Tom, 236
  In Search of Excellence (with Waterman), 62, 151